FLAMMABLE HAZARDOUS MATERIALS

GLENCOE PRESS FIRE SCIENCE SERIES

Bryan: **Fire Suppression and Detection Systems**
Bush/McLaughlin: **Introduction to Fire Science,** Second Edition
Carter: **Arson Investigation**
Clet: **Fire-Related Codes, Laws, and Ordinances**
Erven: **Fire Company Apparatus and Procedures,** Second Edition
Erven: **First Aid and Emergency Rescue**
Erven: **Handbook of Emergency Care and Rescue,** Revised Edition
Erven: **Techniques of Fire Hydraulics**
Gratz: **Fire Department Management: Scope and Method**
Meidl: **Explosive and Toxic Hazardous Materials**
Meidl: **Flammable Hazardous Materials,** Second Edition
Meidl: **Hazardous Materials Handbook**
Robertson: **Introduction to Fire Prevention**

Consultant:

CHIEF DAVID B. GRATZ

Fire Management Associates, Inc.
Silver Spring, Maryland

FLAMMABLE HAZARDOUS MATERIALS

SECOND EDITION

JAMES H. MEIDL
California Fire Service Training Program
California State Department of Education

(A Glencoe Book)

MACMILLAN PUBLISHING CO., INC.
New York
COLLIER MACMILLAN PUBLISHERS
London

Printed in the United States of America

Glencoe Publishing Co., Inc.
17337 Ventura Boulevard
Encino, California 91316
Collier Macmillan Canada, Ltd.

Library of Congress Catalog Card Number: 73-7364

PRINTING 13 14 15 YEAR 3 4 5

ISBN 0-02-476570-8

Contents

Preface

Those who are responsible for protecting human life and property from fire must face the consequences of progress. New industrial materials are being introduced at an ever increasing rate to satisfy the requirements of modern civilization. Many of these materials have properties that make them dangerous. It is the job of fire fighters to recognize and combat these hazards.

The chemical population explosion continues all around us. So much so that the National Fire Protection Association (NFPA), which used to revise its *Hazardous Materials* pamphlet about once a decade, now revises it yearly, just to keep even. Also, the use of large groups of hazardous materials has grown over the past thirty years.

The properties of plastics, combustible metals, cryogenic materials, new and more lethal insecticides, radioactive substances, and rocket propellants are of direct concern to members of the safety services. These substances are part of our modern world, and it is the members of the safety services who must handle them during an emergency.

The problem is more than one of growing numbers. An industrial material can be hazardous in several ways: it may burn or explode, react violently when exposed to air, water, or other chemicals stored nearby; it may be corrosive or poisonous, capable of penetrating even unbroken skin.

Consider water-reactive materials. With such materials, a stream of water may cause a fire or explosion, or tremendously accelerate the burning rate. A sprinkler system can become a liability. We enter a world in which water is an enemy, not a friend. How is such a fire handled? Further, once burning, some materials cannot be extinguished by any known method. Yet, lives have been risked in fighting just such lost causes. To approach certain poisons safely, fire fighters must wear special protective clothing—clothing that is not often immediately available. What can be done then?

To add to the problems, many materials combine hazards: one may be toxic and explosive; another may react violently with water and also be corrosive; still another may be flammable and a skin-penetrating poison. Various labeling and identification systems are designed to help us sort out this bewildering variety of properties, but some of these systems fail to recognize multiple hazards. How can one remember what the hazards are?

As the NFPA handbook points out:

> It is customary to classify a chemical by its predominant (greatest) hazard. The danger of such a procedure, of course, is that the unmentioned hazardous property may be overlooked.

Clearly, one cannot depend entirely on labels and placards. They may be misleading or incomplete.

When we study hazardous materials, we find no hard-and-fast formulas like those found in hydraulics; formulas that, once learned, remain true in all situations. As new data and fire histories come in, opinions change. Much of what we learn today may be obsolete in five years or less. In spite of these complications, we still must learn all we can about hazardous materials because too many lives will depend on our knowledge. We want the answers to such practical questions as: "Which chemicals are dangerous? How do I recognize them? What do I do (and not do) in an emergency?"

In a scientific age, we are also concerned with the why of things, such as: "Why won't this material burn? Why is this metal water-reactive? Why is carbon monoxide toxic?" Those who are not fully convinced of the value of studying information that may rapidly become outdated may ask other questions, too, such as: "My city is mainly bedrooms with just a little industry. When will I need to use my special knowledge of hazardous materials?"

Make no mistake. We are surrounded by hazardous materials. No matter what is burning, toxic gases will be given off. Every house fire fills the interior of the burning building with the gas that is the major cause of fire death: carbon monoxide. Wherever we are, at home, on the highway, or at work, there is flammable liquid not far away. How many service stations or gasoline tanks are there in any city? During the past twenty-five years, the use and misuse of gasoline, kerosene, naphthas, alcohols, paints, and thinners have killed and maimed thousands of people, including many fire fighters. Most fatal explosions are caused by familiar materials—natural gas, flammable liquid vapors, dusts, liquid petroleum (LP) gases, acetylene, and others—materials that are all around us wherever people live.

Furthermore, although a city may not be industrialized, it will have a railroad or a highway running through it. We can be certain that many of the chemicals we will discuss will be on a truck inside your city's limits at some not-too-distant date. We are not even safe from the dangers of radioactivity. Within the past few years, radioactive materials have been spilled in a large city that had special crews trained to handle this hazard; but it has also been spilled in a small town with only a volunteer fire department.

In general, industrial materials present the greatest potential danger to us, wherever we live or work; these are the materials that cause the big fires—the dangerous ones—and tragic loss of life. Often, the reason for this loss of life is ignorance.

Some veteran fire fighters insist it is better not to know all about these hazards. The knowledge, they feel, would keep fire fighters from getting close enough to handle an emergency, since there are many situations for which no clear-cut program of action can be suggested. In a fire situation, with flowing flammable liquids, with mixtures of gases and vapors swirling in the heated air, with leaking or exploding drums of chemicals, nobody can predict accurately what will happen next. These veterans also point out the past mistakes in evaluating hazards of materials such as ammonium nitrate and maintain that, even with all the knowledge we can get, tragedies are still possible.

In spite of these uncertainties, there is still no substitute for knowledge and training if fire fighters are to operate efficiently at a fire involving hazardous materials. There may be nothing else (or nobody else) on hand to help. A label may be inadequate or nonexistent or invisible in the smoke of a fire. The lives of fire fighters should never depend on what a truck driver or security guard might know. Industrial representatives may have no idea of what their stored materials will do if accidentally spilled or involved in a fire. Nor can fire fighters depend on plant managers, for these managers depend on the expertise of police and fire department personnel in an emergency.

It is up to fire fighters to learn what they can if they are to meet their responsibilities. While it is true that a little knowledge of the dangers of chemicals can be confusing, it is equally true that no knowledge at all can be disastrous.

FLAMMABLE HAZARDOUS MATERIALS

1 The Chemistry of Fire

A material may be toxic, reactive, or explosive under certain conditions. But the first question we ask when we are faced with an emergency involving an unknown chemical is, Does it burn? A practical understanding of flammable materials is rooted in the answers to such questions as, "What is combustion?" or "Why are some chemicals flammable and others nonflammable?" These questions in turn cannot be answered without referring to the basic principles of chemistry.

Many of us, through disuse, have forgotten most of our high school or college science courses. In this chapter a refresher will be presented which we hope will prove helpful. Many of the terms we will discuss are found frequently in the literature about hazardous materials.

A good introduction to questions about flammability is a discussion of the FIRE RECTANGLE (Figure 1–1). Most of us are aware that a combination of

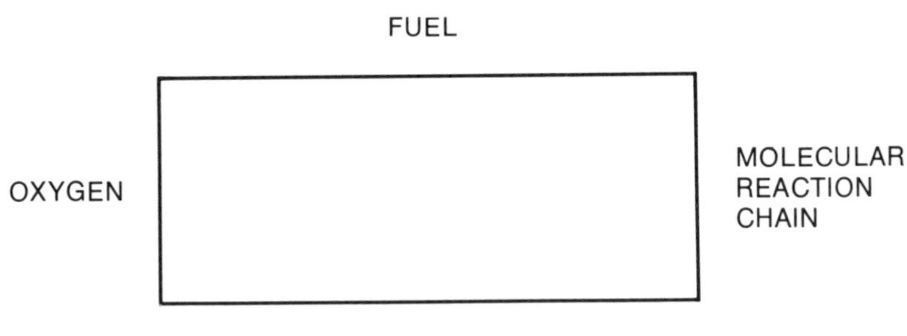

FIGURE 1–1. The Fire Rectangle

three factors—FUEL, HEAT, and OXYGEN—is necessary before combustion can occur. This is the famous FIRE TRIANGLE. However, in recent years it has been discovered that a fourth factor is involved, the MOLECULAR REACTION CHAIN. Fires can be extinguished with dry chemical powders, carbon tetrachloride, and the chlorinated hydrocarbon gases such as the Halons because application of these substances breaks the molecular reaction chain during combustion.

Later in this chapter we will examine each of the four factors in the fire rectangle in some detail. But, before we do, we must learn the meaning of some basic chemical terms and concepts.

Basic Chemistry

Atoms, Elements, Molecules, Compounds, and Mixtures

Basically, all matter is made up of ATOMS, which are the single units or "building blocks" of everything that occupies space and has weight. There are 106 different kinds of atoms thus far identified, each of which is referred to as an ELEMENT.

Elements vary widely in their properties. Most are metals. Most are solids. Some are gases. Bromine, a nonmetal, and mercury, a metal, are liquids at room temperature. As far as we know, the entire universe is made up of less than one hundred of these elements, some of which are extremely rare. The other elements are artificially produced in the laboratory. Only 88 occur naturally on our earth, and only about 30 elements are at all common. Yet, these few "building blocks" form literally billions of combinations, because molecules can contain from two to over a dozen different kinds of atoms in different combinations.

A MOLECULE is a single unit of two or more atoms that have chemically combined. Some elements, such as fluorine, chlorine, bromine, and iodine, combine in pairs with themselves. These are called diatomic molecules.

A second kind of molecule is a single unit of two or more *different kinds* of elements chemically combined. These are COMPOUNDS. An atom is the smallest unit that can exist of an element. A molecule is the smallest unit that can exist either of the diatomic elements or of a compound.

Compounds do not resemble the elements from which they are formed. A good example is common table salt, chemically known as sodium chloride. It is made of two elements, sodium and chlorine. Sodium is a dangerous solid element which will explode on contact with water; chlorine is a poisonous, greenish-yellow gas. When chemically combined, the sodium chloride molecules form the familiar salt crystals known to all of us.

Both atoms and molecules are extremely small. In a glass of water there are more water molecules than you could ever count in your entire lifetime.

TABLE 1–1. MELTING AND BOILING POINTS OF SOME ELEMENTS

SOLID ▬ LIQUID ▒ GAS ░

DEG. F. DEG. C.	0 −18	1000 538	2000 1093	3000 1649	4000 2204	5000 2760	6000 3316
Aluminum							
Bismuth							
Bromine							
Chlorine							
Copper							
Gallium							
Iodine							
Iron							
Lead							
Magnesium							
Mercury							
Nitrogen							
Oxygen							
Phosphorus							
Sodium							
Sulfur							
Tin							
Zinc							

In just a teaspoonful of the element sulfur, there are also more individual sulfur atoms than you could ever count.

MIXTURES are variable compositions of elements, compounds, or both that have not combined chemically. The components of a mixture do not lose their individual identities, and they can be separated by physical or mechanical means.

Atomic Symbols

In order to show what elements or combinations of elements are present in a substance, it is both convenient and necessary to learn a sort of chemical shorthand. Each element is represented by either a single letter or no more than two letters. C stands for carbon; H, hydrogen; O, oxygen; P, phosphorus. Two-letter symbols include Cl for chlorine; He, helium; Li, lithium; Ne, neon. When one letter is used, it is always capitalized. When two letters are used, the first letter is capitalized but the second letter is not.

Some symbols of elements do not seem to match their names at all. This is because some of the symbols come from ancient Latin names: Fe, iron, from *ferrum;* Au, gold, from *aurum;* Pb, lead, from *plumbum* (from which we get the word *plumber,* by the way); and K, potassium, from *kalium.* Through repeated use these symbols will become familiar to you, just as you substitute *pound* in your mind when the sign reads *lb.*

TABLE 1–2. SOME COMMON ELEMENTS

ELEMENT	SYMBOL	APPROXIMATE ATOMIC WEIGHT
Aluminum	Al	27
Antimony	Sb	122
Arsenic	As	75
Barium	Ba	137
Bismuth	Bi	209
Boron	B	11
Bromine	Br	80
Cadmium	Cd	112
Calcium	Ca	40
Carbon	C	12
Chlorine	Cl	35.5
Chromium	Cr	52
Cobalt	Co	59
Copper	Cu	63.5
Fluorine	F	19
Gold	Au	197
Hydrogen	H	1
Iodine	I	127
Iron	Fe	56
Lead	Pb	207
Magnesium	Mg	24
Manganese	Mn	55
Mercury	Hg	201
Nickel	Ni	59
Nitrogen	N	14
Oxygen	O	16
Phosphorus	P	31
Platinum	Pt	195
Potassium	K	39
Silicon	Si	28
Silver	Ag	108
Strontium	Sr	88
Sulfur	S	32
Tin	Sn	119
Titanium	Ti	48
Tungsten	W	184
Uranium	U	238
Vanadium	V	51
Zinc	Zn	65

Table 1–2 lists the chemical symbols and atomic weights of forty of the more common elements. It would be to your advantage to commit them to memory.

Chemical Formulas

Once you have memorized the symbols of the elements you are on your way. Yet, there is still more to memorize if you are to be able to read chemical names. In nature, certain chemical combinations occur over and over again: sulfates, nitrates, carbonates, and acetates, just to name a few. Without going very deeply into the system of chemical nomenclature, we have listed in Table 1–3 some of the chemical symbol combinations that you are sure to run across.

In chemical formulas, the first symbol is generally read first:

K, potassium
Na, sodium
Al, aluminum
Pb, lead

The second symbol or group of symbols is then read:

NO_3, nitrate
Cl, chloride
CO_3, carbonate
SO_4, sulfate

KNO_3, therefore, is read as *potassium nitrate*. NaCl is sodium chloride. $CaCO_3$ is calcium carbonate. $BaSO_4$ is barium sulfate.

TABLE 1–3. COMMON CHEMICAL COMBINATIONS

CHEMICAL FORMULA ENDING	NAME OF SUBSTANCE
$-C_2H_3O_2$	acetate
$-HCO_3$	bicarbonate
$-BO_4$	borate
$-Br$	bromide
$-CO_3$	carbonate
$-Cl$	chloride
$-CrO_4$	chromate
$-CN$	cyanide
$-F$	fluoride
$-OH$	hydroxide
$-I$	iodide
$-NO_3$	nitrate
$-O$	oxide
$-PO_4$	phosphate
$-S$	sulfide
$-SO_4$	sulfate

TABLE 1–4. COMMON NAMES OF CHEMICAL COMPOUNDS

COMMON NAME	CHEMICAL NAME	CHEMICAL FORMULA
Baking soda	Sodium bicarbonate	$NaHCO_3$
Bluestone	Cupric sulfate	$CuSO_4 \cdot 5H_2O$
Brimstone	Sulfur	S
Grain alcohol	Ethanol, or Ethyl alcohol	C_2H_5OH
Lime	Calcium oxide	CaO
Limestone	Calcium carbonate	$CaCO_3$
Litharge	Lead oxide	PbO
Lye	Sodium hydroxide	NaOH
Milk of magnesia	Magnesium hydroxide	$Mg(OH)_2$
Muriatic acid	Hydrochloric acid (usually a "commercial" grade)	HCl
Oil of vitriol	Sulfuric acid	H_2SO_4
Plaster of paris	Calcium sulfate	$CaSO_4 \cdot \frac{1}{2}H_2O$
Potash	Potassium carbonate	K_2CO_3
Rubbing alcohol	Isopropyl alcohol	$CH_3CHOHCH_3$
Wood alcohol	Methanol, or Methyl alcohol	CH_3OH

This is how it usually works—but not always. Chemists, and especially those who are not chemists, may refer to substances by their "common names," which often date back to ancient times. Table 1–4 lists the common, household names for some chemical compounds.

It is a good idea to remember these—and even to add to the list. If you are on the fire dispatching board some evening and a frantic voice tells you over the telephone, "My child has just swallowed some lye!" or "My husband has just spilled muriatic acid on his hands!" what do you do? What should you tell the person who calls?

Structural Formulas

Unfortunately for the chemistry student, there are other methods of chemical formulation or nomenclature. Since some of these methods are widely used, we will have to familiarize ourselves with them.

Butane, the common LP (liquid petroleum) gas, contains four carbon atoms and ten hydrogen atoms in its molecule. Its molecular formula is C_4H_{10}. But, for some purposes, this is not enough. We may wish to show the exact relationship of the various atoms in the butane molecule through the use of the STRUCTURAL FORMULA (Figure 1–2).

We can easily count the atoms—four carbons and ten hydrogens—but we can also see how the carbons are linked to the hydrogens. This pattern is represented by the formula, $CH_3CH_2CH_2CH_3$. In another form, the chemist writes, $CH_3(CH_2)_2CH_3$. By putting part of the formula in parentheses, the chemist tells us to repeat this part; and the number beside the parenthesis, called a subscript, tells how many times the part should be repeated.

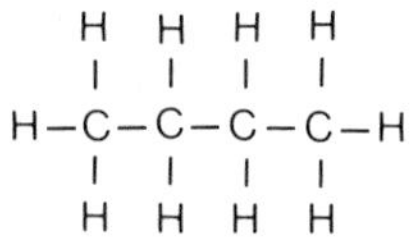

```
    H   H   H   H
    |   |   |   |
H — C — C — C — C — H
    |   |   |   |
    H   H   H   H
```

FIGURE 1–2. The Butane Molecule (C_4H_{10})

Process Symbols

Often, in describing the way two different compounds combine, arrows show the way the process proceeds. An arrow pointing upwards $\uparrow$ shows formation of a gas. An arrow pointing downward $\downarrow$ symbolizes formation of a solid, often called a *precipitate*. Used at the center of an equation, an arrow points out the direction of the reaction $\rightarrow$. Arrows pointing in opposite directions $\rightleftarrows$ show that the process is reversible.

Constructing an Equation

Let us now take the symbols we have discussed and use them to construct an equation showing the combination of two compounds. Acetylene is produced commercially by combining calcium carbide, CaC_2, with two water molecules, 2 H_2O. This combination produces acetylene gas, C_2H_2, and a residue of slaked lime, $Ca(OH)_2$. We can describe all this simply by writing, $CaC_2 + 2\ H_2O \rightarrow Ca(OH)_2 + C_2H_2\uparrow$.

Notice that all the atoms are accounted for in this equation. There are the same number of atoms of the same kind on each side of the center arrow. Count them and see. Remember: On the molecular level, whether in a reaction of this kind or in combustion, every atom remains the same although it may shift position and join with one or more atoms to form a new molecule. Later, we will see that atoms themselves can change only at the nuclear level when their interior structure is altered.

Fire Rectangle

Fuels

Now that we have learned something about the nature of elements and molecules, let us discuss fuels. Fuels are those substances that will burn when heat is applied to them. Carbon, hydrogen, sulfur, magnesium, titanium, and some other metals are examples of pure elements that can burn. Coal, charcoal, and coke, for example, are almost pure carbon; hydrogen, another element, is a highly flammable gas. But the most familiar combustible materials are not pure elements; they are compounds and mixtures.

Wood, paper, and grass are composed principally of molecules of cellulose, a flammable substance. If we examine the chemical makeup of this compound, we will discover what elements form the basic fuels in most solid materials. The cellulose molecule contains twenty-one atoms: six carbons, ten hydrogens and five oxygen atoms: $C_6H_{10}O_5$. Since oxygen is not flammable (see OXYGEN, below), it follows that the carbon and hydrogen found in most common combustible solids are the elements that burn. This conclusion becomes more certain if we look at common flammable liquids. Gasoline, kerosene, fuel oils, and other petroleum compounds are composed of only carbon and hydrogen atoms, in varying amounts. These compounds, called HYDROCARBONS (hydrogen + carbon), will all burn.

Other flammable compounds are composed of carbon, hydrogen, and oxygen atoms in a fixed ratio, making it appear as if there is a water molecule attached to each carbon atom. A good example is glucose, a common sugar, which has the formula $C_6H_{12}O_6$. Chemists call this type of molecule a "hydrated (watered) carbon," or carbohydrate. Carbohydrates also burn but are not to be confused with hydrocarbons.

Carbon and hydrogen are only two of the elements which will burn. But, since most common flammable materials contain a combination of carbon and hydrogen fuels, we will limit our discussion of combustion to them at this point.

Fuel, as we have seen, is only one side of the fire rectangle. Before it will burn, any fuel requires the addition of heat, another side of the rectangle.

Heat

The fire fighter's work is directly related to the basic principles of heat. Dissipating unwanted heat (fire) is one way in which we make our living. The generation of heat causes many of our problems.

Heat Energy Heat energy (our main concern in this book) can be produced by the building up of molecules—composition and polymerization; by the breaking apart of molecules—decomposition; by heat of solution when materials are dissolved in a liquid; or by combustion—another form of composition. Energy is defined as the ability to do work.

MECHANICAL HEAT ENERGY can be produced by friction (a common cause of fire), the compression of a gas, or by applying the brakes on an automobile.

ELECTRICAL HEAT ENERGY can be produced by lightning, arcing, the discharge of static electricity, electrical resistance, etc.

NUCLEAR HEAT ENERGY can be produced by fission (in atom bombs), fusion (in the hydrogen bomb), or a combination of both. Nuclear energy is created on the sun by nuclear fusion.

Fission is the splitting apart of very heavy atoms such as Uranium-235 into lighter atoms with a resulting loss of mass or weight and the conversion of this lost substance into energy.

Fusion is the ramming together of very light atoms such as some forms of hydrogen (deuterium and tritium), producing heavier atoms, and again releasing tremendous amounts of energy.

Heat Transfer A law of physics states that heat tends to flow from a hot substance or place to a cold substance or place. This fact explains the ability of a material to absorb heat from another. Once produced, heat is transferred (from hot to cold) in three ways.

CONDUCTION is the transfer of heat *through a medium*. Metals are the best conductors of heat. As an example, a pot which is heated by the stove might conduct the heat to the hand of the person who picks it up.

CONVECTION is the transfer of heat *with a medium*. This type of heating is usually circulatory; the medium becomes heated, moves to a cooler spot carrying its heat with it, gradually cools off, and returns to its original location. The process is repeated until all the heat is dissipated. The hot air and gases that rise in a multi-story-building fire carry heat by convection and start fires on the upper floors. In nuclear reactors molten sodium is used to transfer the heat by convection from one part of the reactor to another.

RADIATION is the transfer of heat which is *not dependent upon any medium*. The heat which travels from the sun to the earth through a vacuum (no medium) is a good illustration of this method. Glass, which is a poor conductor of heat, passes radiated heat quite readily.

Heat vs. Temperature As we have just seen, heat can be produced in four ways and transferred in three. We come now to the question, What is heat? *Heat is the total amount of vibration in a group of molecules.*

Molecules, either as elements or compounds, vibrate constantly. If the molecules in a material vibrate slowly enough to remain locked in position, we call the material a solid. As energy is applied, the vibratory rate of the molecules increases. Finally, the fixed position cannot be maintained. Molecules begin to slide over each other, although still remaining in contact, and the solid becomes a liquid. We say the material has melted. As additional energy is applied, the molecules in the liquid begin to vibrate so rapidly that they cannot remain close together. They fly off in all directions; the liquid boils and becomes a gas. This conversion of solids and liquids into gases is an important part of combustion.

What we call TEMPERATURE is simply a measure of how fast molecules are vibrating. A "hot" material is one with a high vibratory rate. We burn our fingers when we touch a hot stove: contact with a hot material causes the skin and flesh molecules to start vibrating at a high rate which damages their delicate structure and causes pain.

Note: There is a difference between TEMPERATURE and HEAT. Temperature is a measure of *the vibratory rate* of a molecule; heat is a measure of the energy given off by the *total amount of vibration* in a group of molecules. For example, a drop of boiling water has a high temperature because all its molecules are vibrating rapidly. A sink full of lukewarm water actually has much more heat than the drop of water because it has far more molecules vibrating, even though they are not vibrating as rapidly.

Temperature Scales There are two temperature scales widely used throughout the world today (see Table 1–5), the Fahrenheit scale (in the English system of weights and measures) and the Celsius scale (in the metric system of weights and measures). The English system is used only in the United States, Canada, and Great Britain. All of these countries are presently in a period of transition from the old and cumbersome English system to the metric system. Therefore, both systems of weights and measures will be used in this text.

TABLE 1–5. SOME COMMON TEMPERATURE SCALE COMPARISONS

	CELSIUS	FAHRENHEIT
Boiling point of water	100°	212°
Normal human body temperature	37°	98.6°
Normal room temperature	20°	68°
Freezing point of water	0°	32°

To change from Fahrenheit to Celsius, a conversion formula can be used. Just take the number of degrees Fahrenheit, subtract 32 from this figure, and multiply the result by five-ninths.

$$°C. = 5/9(°F. - 32)$$

To change from Celsius to Fahrenheit, multiply the number of degrees Celsius by nine-fifths and then add 32 to the answer.

$$°F. = (9/5)°C. + 32$$

Melting and Boiling Points Almost any group of molecules, at some given temperature and pressure, can exist as a solid, a liquid, or a gas (see Table 1–1). Because their physical structure varies, elements and compounds melt or boil at different rates of vibration. Oxygen, for example, vibrates rapidly enough to melt at −361° Fahrenheit; it boils, to form a gas, at −297°F. (−182.7°C.) A water molecule requires a much higher vibratory rate to melt—just over 32°F. (0°C.)—and it boils at 212°F. (100°C.). Iron requires a tremendous rate of vibration to melt: 2795°F. (1535°C.).

Oxygen

We come now to the third side of the fire rectangle, OXYGEN, the most common element on earth. Oxygen makes up a major portion of the oceans and of the earth's crust, and one-fifth of our atmosphere. ATMOSPHERIC OXYGEN is the major source of the oxygen that supports combustion. While other elements can support combustion—chlorine and fluorine are two examples—most fires are aided by oxygen.

Oxygen itself does not burn. An electric arc or other spark will not explode any given concentration of oxygen. A match flame will ignite or explode a flammable gas but the match itself burns faster when brought in contact with oxygen.

Oxygen supports combustion. Indeed, without oxygen (or some substitute) combustion is impossible. Normal burning is the combination of fuels with oxygen under the influence of heat. It is important, however, for the fire fighter to remember that certain flammable chemicals have oxygen built into their molecules which is available for combustion even when there is no atmospheric oxygen available. Other materials, called OXIDIZING AGENTS, *release oxygen when heated.* Although they may be nonflammable themselves, chemicals that release oxygen make any fire in nearby materials more intense.

Oxidation vs. Combustion Oxidation is not necessarily combustion. Human beings take in about 200 gallons of atmospheric oxygen each day. We generate enough heat by the oxidation of fuels (food) to maintain our bodies at 98.6°F. (37°C.). Rust, the slow combination of iron with oxygen, is a form of oxidation. So are corrosion, the yellowing of the pages of old books and magazines, and certain explosions. Each of these represents a different rate of oxidation.

Combustion Simply defined, *combustion is a rapid oxidation or chemical combination accompanied by heat, and usually, light.* With this in mind, we can look at the fire rectangle and see how its four parts combine.

$$C + O_2 \rightarrow CO_2$$

Carbon and oxygen unite to produce carbon dioxide.

$$2C + O_2 \rightarrow 2CO$$

Carbon and oxygen unite to produce carbon monoxide.

$$2H_2 + O_2 \rightarrow 2H_2O$$

Hydrogen and oxygen unite to produce water.

Oxygen is not always necessary to have oxidation. Chlorine gas does not burn but, like oxygen, will support combustion.

$$Cu + Cl_2 \rightarrow CuCl_2$$

Copper metal and chlorine unite to produce cupric chloride.

A molecule of elemental oxygen contains two oxygen atoms, O_2. When a carbon atom combines with oxygen, it forms a combustion product, carbon dioxide, CO_2, and energy is released. If there is insufficient oxygen in the air, two carbon atoms can combine with one oxygen molecule. The combustion product is two molecules of carbon monoxide, 2CO. Again, there is a release of energy.

A hydrogen molecule, our other fuel, contains two atoms, H_2. When two hydrogen molecules are oxidized, the combustion product is two molecules of water vapor, $2H_2O$.* Again, energy is released. The energy released by these combinations comes from the oxygen molecule. There is more energy in the oxygen molecule than in the carbon dioxide, carbon monoxide, or water molecules. When the oxygen molecule breaks up and reforms into carbon dioxide or water, this excess energy is released. The breakup of the oxygen molecule is generally caused by the influence of heat.

Spontaneous Ignition Picture a pile of rags soaked in linseed oil. The linseed oil molecules are constantly being oxidized by the air. This combination releases energy. If the rags are tightly baled or if they are stored in a tightly closed metal container, there is not enough oxygen available for combustion. The formation of heat is held down. But, in a loose pile of oily rags, enough air will reach the interior to produce heat. And, at the same time, the insulating effect of the pile itself retards the dissipation of this heat. The temperature begins to rise, the molecules become more energetic (they vibrate faster) and they combine with oxygen at an increasing rate. This raises the temperature still further. Finally, the reaction moves along so rapidly that it produces visible heat (flames).

This process is called SPONTANEOUS IGNITION. Spontaneous ignition of materials like oily rags and agricultural products (where the initial heat rise is partly caused by bacterial action) *may cause a fire.*

Open Flames When the heat of a burning match is applied to a piece of paper, the cellulose molecules of the paper start to vibrate at an enormously increased rate. And, almost instantaneously, they begin to break apart. In a series of reactions, these fragments continue to break up, producing the free carbon and hydrogen that will combine with oxygen. This combination releases additional energy. Some of the released energy breaks up still more cellulose molecules, releasing more free carbon and hydrogen, which, in turn, combine with more oxygen, releasing more energy, and so on.

We have here a reaction chain which will continue until the fuel is exhausted, the oxygen is excluded in some way, the heat is dissipated, or the reaction chain is broken. The energy from a single match is enough to begin

---

*When the Hindenburg, last of the great German dirigibles, burned at Lakehurst, N.J., in 1937, an intense rainstorm, caused by the burning of the hydrogen used for lift, saved many people trying to escape from the gondola.

a chain which will burn an entire forest. In cellulose, the process looks like this:

$$C_6H_{10}O_5 + 6O_2 \rightarrow 5H_2O + 6CO_2 \uparrow \text{ and heat energy.}$$

Briefly, the heat of the match vaporizes the fuel, cracks it into more easily oxidized fragments, and creates a temperature level high enough to greatly increase oxidation. Cellulose combines with atmospheric oxygen many times faster at fire temperatures than at room temperatures. This is because the speed of a chemical reaction doubles with every 17-degree rise in Fahrenheit temperature (9.5 degrees Celsius), a geometrical progression.

In our example, we assumed an ample supply of air (open flame conditions). It is important to remember, however, that carbon is relatively slow burning and there is often not enough air available to oxidize it all into carbon dioxide. As a result, some carbon leaves the combustion zone as black smoke and floating particles, some remains as charcoal, continuing to burn without visible flame, and some forms carbon monoxide, a toxic and flammable gas. Incomplete combustion, producing large amounts of carbon monoxide, is characteristic of automobile and truck engines and makes the inhalation of exhaust fumes dangerous, if not fatal.

Heat of Combustion As we have seen, burning fuels produce heat energy. Since each fuel has a particular arrangement of atoms, each burns at a particular rate which is different from any other. The amount of heat a fuel releases during its complete oxidation is called its HEAT OF COMBUSTION. Heats of combustion of some common fuels, expressed in British Thermal Units (BTU)* released per pound of material, are as follows:

Paper, Cardboard	6,000
Paper, Rags, Wood	7,000
Newspaper, Sawdust, Fibers	8,000
Coal	12,500
Flammable Liquids	16,000 to 21,000
Flammable Gases	20,000 to 23,000
Hydrogen	60,000

While these figures are only approximations, they give a good idea of the relative amounts of heat produced by various types of fuel.

Fire fighters, who are familiar with tests of burning flammable liquids in the open air and with highway fires involving automobile gasoline tanks, may tend to underestimate their heats of combustion. Furthermore, the *rate*

*BTU, the amount of heat required to raise the temperature of one pound of water one degree Fahrenheit, e.g. from 60° to 61°F. Its metric equivalent is the calorie, which is defined as the amount of heat energy necessary to raise one gram of water one degree Celsius. 1 BTU = 242 calories.

of heat climb in a flammable liquid or a flammable gas fire is much faster than in a wood or paper fire. After only a few minutes, such a fire can double its heat, making the interior of a building impossible to enter and difficult to approach. This is one reason why fires involving flammable liquids and gases present special problems.

Summary

We have seen that a fire requires fuel, heat, and oxygen. In addition, it results from a series of reactions in which complicated molecules "crack" into easily oxidized fragments. Disruption of this reaction chain, along with the removal of fuel, heat, or oxygen, is now a recognized method of fire extinguishment.

The Chemistry of Extinguishing Fire

Removal of Fuel

The simplest way to remove fuel is by consumption. When a building burns to its foundations, there is nothing left to burn and the fire goes out. Happily, there are better methods of extinguishment, such as:

1. Setting backfires.
2. Removing threatened combustibles from the fire area.
3. Draining a flammable liquid tank.
4. Shutting off a gas flow.

Removal of Heat

The most common method of fire extinguishment is to remove heat by applying water. It has sometimes been said, "If water didn't exist, fire fighters would have to invent it." But, while water is an almost ideal extinguishing agent in many ways, it does have properties which limit its effectiveness:

1. *Surface tension,* which limits its ability to penetrate tightly packed combustibles.
2. *Conducts electricity*
3. *Reacts with many materials,* which causes the evolution of heat, creation of toxic or flammable gases, and even explosions.
4. *Freezes* at the inconveniently high temperature of 32°F. (0°C.).

5. *Low viscosity,* which means that it runs off quickly when applied to any surface.

6. *Transparency,* which allows radiant heat to pass with little difficulty.

7. *Relatively ineffective* on many flammable liquid fires.

Various additives, including foams, have been designed to overcome some of these limitations. With all its faults, however, water is a tremendously versatile fire-fighting tool. It can be used to direct flammable liquids away from a fire. It can protect personnel and exposures. It can be used to extinguish fires by blanketing, displacement, dilution, and emulsification, as we shall see when we discuss flammable liquids. Nevertheless, *the primary way water puts out fire is through heat removal.*

Specific Heat SPECIFIC HEAT is defined as the ratio between the amount of heat necessary to raise the temperature of a material and the amount of heat necessary to raise the temperature of the same weight of water by the same number of degrees. In effect, specific heat is a comparison. Water is chosen as the standard because it is convenient. It takes more heat to raise the temperature of water than it does to raise the temperature of almost any other substance. Recall that one BTU is the amount of heat required to raise one pound of water by one degree Fahrenheit. Applying this to the definition above, the ratio of one BTU to one BTU is one (1.0), which is the specific heat of water. Since the amount of heat required to raise the temperature of one pound of almost any other substance will be less than one BTU, it follows that the specific heats of most other substances will be less than one.

Water as a Heat Remover The fact that it takes more heat to raise the temperature of water than it does to raise the temperature of almost anything else is very important to fire fighters because it means that when water is applied to any burning material, the temperature of the material will go down faster than the temperature of the water goes up. It also means that much less water will be required to lower the temperature of a given amount of material. If, for example, some substance were used in place of water to put out a fire, and that substance had a specific heat of only .5, twice as much of the substance would be required to lower the temperature of the burning material the same number of degrees.

Suppose we were to mix a gallon of water at 50°F. (10°C.) with a gallon of acetone at 100°F. (37.8°C.). We would expect the resulting mixture to have a temperature of 75°F. (23.9°C.). But water has a much higher specific heat than acetone, 1.0 to 0.506. It takes more energy to heat water 25 degrees than acetone loses in cooling 25 degrees. As a result, the mixture will have a temperature of less than 70°F. (21.1°C.). On a limited scale, this fact can be very useful in dealing with some flammable liquid spills, since these liquids will not burn if cooled below a certain temperature. (See The Dilution Technique, Chapter 2, page 33.)

Latent Heat The LATENT HEAT OF VAPORIZATION is the amount of heat a material absorbs when it changes from a liquid to a gas. Water's high latent heat of vaporization gives it much of its fire-fighting effectiveness. Since we rarely shovel snow or ice onto a fire, LATENT HEAT OF FUSION, the amount of heat a material absorbs when it changes from a solid to a liquid, is of limited importance to the fire fighter. In either case, however, water has an advantage: its latent heats are substantially higher than those of other substances.*

Only 180 BTUs are required to raise the temperature of one pound of water 180 degrees, from 32°F. (O°C.), its melting point, to 212°F. (100°C.), its boiling point. This same pound of water, as it turns to steam at 212°F., absorbs no less than 970 BTUs. An enormous amount of heat energy is absorbed without *any equivalent temperature rise*. This energy is all used to turn sliding molecules into flying molecules.

Because of water's high latent heat of vaporization, we use a fog pattern of fine droplets to get a maximum cooling effect: fog is a form of water that turns readily into steam. Note: Steam is invisible. The heat absorption process ends when steam condenses into visible water droplets; condensation of water vapor above a fire has no cooling effect on burning materials.

When water turns to steam, it also increases in volume some 1700 times. Sometimes, if enough steam is generated, air can be displaced around a fire, and extinguishment from oxygen starvation then follows. But, steam is not a major factor in extinguishing ordinary fires. As a matter of fact, it can even cause a fire. Even though it is generated at the boiling point of water, 212°F. (100°C.), steam temperature rises several hundred degrees higher than that—high enough to ignite many combustible materials.

An attempt to use steam as an extinguishing agent may have contributed to the great fire disaster at Texas City. Remember also, steam is useless against oxidizing materials. (See Chapter 11.) *The ability to absorb heat and to generate steam is what makes water such an effective extinguisher.*

Removal of Oxygen

For a fire to be sustained in ordinary combustibles, its fuel fragments must combine with oxygen molecules. If the oxygen concentration falls below a certain level, the fire goes out. When we cover a burning pan of grease,

*The heat given off when water turns to ice, or steam turns to water, is also called *latent heat*.

shovel dirt on smoldering grass, use foam or carbon dioxide, or recommend an inert atmosphere of nitrogen or carbon dioxide above a flammable liquid storage, we are attacking or preventing fire by removing the oxygen supply; in effect, we are shutting off the atmosphere. The same thing occurs in a cellar or other enclosed space. Once the combustion process uses up the oxygen available, the fire is reduced to smoldering until a fresh supply of oxygen is introduced.

Since 21 percent of the earth's atmosphere is oxygen, air is the primary source of oxygen. Fortunately, the remaining 79 percent of the air is almost entirely made up of inert gases that neither burn nor support combustion. Otherwise, the first fire ever started would have become worldwide in short order. Shutting off the air supply will extinguish most fires, but, as we have already mentioned, oxidizing agents can accelerate a fire by releasing oxygen. Nitrates and chlorates, two such oxidizing agents, contain three available oxygen atoms in their molecules. This immediately raises the question, Why is oxygen available in some molecules, like the nitrates, and unavailable in others? Carbon dioxide is made up of a burnable carbon atom and two oxygen atoms. Why, with a potential like this, is CO_2 so inert that it can be used as an extinguisher? For the answers, we must touch briefly on a very complex subject, *chemical valence*.

Chemical Valence

The valence of any atom determines how it will combine with other atoms. We may visualize a valence as an "invisible electric hook." Valences of some common atoms are:

Hydrogen, chlorine	1
Oxygen	2
Nitrogen	3
Carbon, silicon	4

In Figure 1–2, we showed a butane molecule, composed of four carbon atoms and ten hydrogen atoms: C_4H_{10}. Why this ratio? Why four carbon atoms to ten hydrogens? Why not six hydrogens, or twelve?

The answer is that every atom generally combines with a certain number of other atoms because of electrical attraction. Carbon, with a valence of four, will combine with four hydrogen atoms (hydrogen has a valence of one) to form molecules of methane, the common cooking gas, or with four chlorine atoms (chlorine also has a valence of one) to form molecules of carbon tetrachloride. The carbon atom acts as if it had four "hooks" ready to pick up other atoms (Figure 1–3, page 18).

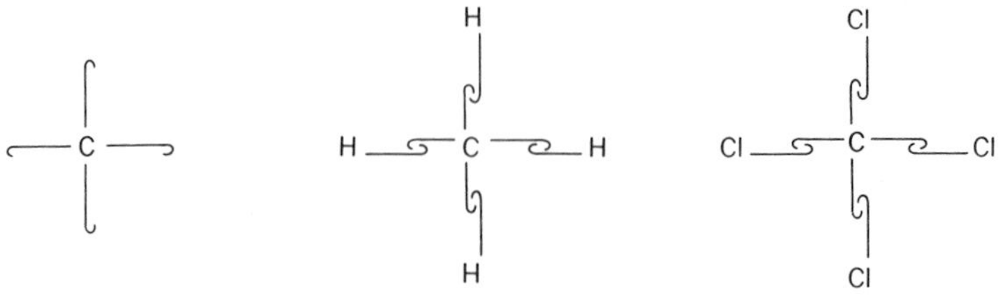

The carbon atom **Methane molecule CH_4** **Carbon tetrachloride molecule CCl_4**

FIGURE 1–3. Valence of Carbon

Similarly, nitrogen, generally having a valence of three, will combine with three hydrogen atoms to form a molecule of ammonia. Oxygen, having a valence of two, will combine with two hydrogen atoms to form water. Hydrogen atoms (valence of one) are seldom found free in nature. Most of the time, they satisfy their valence by joining with other hydrogen atoms to form hydrogen molecules. When they do form compounds with other elements whose valence is one, these compounds look like the molecule of hydrogen chloride.* See Figure 1–4 for examples.

Carbon has a valence of four; oxygen, a valence of two. When carbon and oxygen combine, each oxygen atom *satisfies* two of the valences of carbon. Two *double bonds* are formed and two carbon dioxide molecules result. Silicon dioxide, a major component of bricks, is a molecule similar to carbon dioxide.

What has happened is that all the valences of the atoms in a carbon dioxide molecule have been satisfied. The carbon atom, to put it another way, has been fully oxidized (see Figure 1–5). It cannot combine with additional oxygen; it is completely "burned" already. This is why CO_2 is generally inert and nonflammable and why bricks of silicon dioxide can be used to line furnaces.

A carbon *monoxide* molecule (Figure 1–6), however, shows us something different; carbon monoxide has not reached this satisfied state of full oxidation. Two valences were left unused when the molecule formed because there was not enough oxygen available. Therefore, carbon monoxide is a flammable gas.

*When chemists draw structural formulas, they allow a single line to stand for the connection between atoms.

pressure not exceeding 40 pounds per square inch—2069 torrs*—at 100°F. (38°C.).

A COMBUSTIBLE LIQUID is any liquid whose flash point is 100°F. (38°C.) or above.

Flammable liquids and combustible liquids are subdivided into classes, as shown in Tables 2–1 and 2–2.

Do not worry about the meaning of the terms *flash point* and *vapor pressure*; they will be explained later in this chapter. Tables 2–1 and 2–2 and the definitions above are included here to stress the nature of the materials with which we are dealing.

TABLE 2–1. CLASSES OF FLAMMABLE LIQUIDS

CLASS	FLASH POINT	BOILING POINT
IA	Below 73°F. (23°C.)	Below 100°F. (38°C.)
IB	Below 73°F. (23°C.)	At or above 100°F. (38°C.)
IC	At or above 73°F. (23°C.) and below 100°F. (38°C.)	

TABLE 2–2. CLASSES OF COMBUSTIBLE LIQUIDS

CLASS	FLASH POINT
II	At or above 100°F. (38°C.) and below 140°F. (60°C.)
IIIA	At or above 140°F. (60°C.) and below 200°F. (93°C.)
IIIB	At or above 200°F. (93°C.)

As the quantity and the uses of flammable liquids continue to increase, so do the opportunities for accidents. These liquids cause big fires because they are often stored in bulk, in containers and tanks that hold millions of gallons. A single mistake can cause a fire that lasts for days. In addition, the way flammable liquids act in a fire, because of their physical properties, contributes greatly to the fire spread.

*A TORR is a unit of pressure in the metric system. Torricelli, a seventeenth-century Italian scientist, found that atmospheric pressure (14.7 pounds per square inch at sea level) could be measured according to its ability to support a column of mercury in a glass tube. This discovery led Torricelli to invent the first barometer. The torr, named in his honor, is a unit equal to the pressure that will support a column of mercury one millimeter high. 14.7 pounds per square inch (also known as 1 atmosphere of pressure) will support 760 millimeters of mercury, or 760 torrs.

Physical Properties and Hazards

Although there are thousands of different flammable liquids, the number of physical properties that are important to the safety services are few. Failure to understand the properties of a liquid implies lack of awareness of its potential danger and behavior.

Table 2–3 lists the properties of eight common flammable liquids. To someone unfamiliar with the headings, FLASH POINT, IGNITION TEMPERATURE, FLAMMABLE LIMITS, SPECIFIC GRAVITY, VAPOR DENSITY, BOILING POINT and WATER SOLUBILITY, Table 2–3 has little meaning. To someone who knows how to interpret it, the table tells how a particular liquid will act in a fire, under what conditions it will ignite or explode, and how to handle it in an emergency. The seven key terms appearing above the table columns explain the hazards of thousands of liquids that may differ completely in their uses. Let us see what these key terms mean.

Flash Point

FLASH POINT, the temperature at which a flammable liquid produces enough vapors to be ignited, is the most significant property of all. More than any other single factor, flash point determines the flammability of a liquid. Every hazard rating of flammable liquids recognizes this fact. The reason flash point is so important is that *flammable liquids are not ordinarily flammable. It is the vapors they produce* that ignite or explode.

Gasoline (see Table 2–3) has flash points which vary from –36°F. (–38°C.) to –45°F. (–43°C.). Unless we are wintering in central Siberia, gasoline will ignite when we supply a source of ignition. Kerosene, by contrast, has a flash point of over 100°F. (38°C.). Kerosene will not burn until the liquid is heated to that temperature; for only then will it release enough vapors to ignite. At normal atmospheric temperatures, gasoline is always ready to ignite; this is why its misuse is so much more dangerous than the misuse of kerosene.

The NFPA defines flash point as "the lowest temperature of the liquid at which it gives off vapor sufficient to form an ignitable mixture with the air near the surface of the liquid or within the vessel used."

Notice that this definition uses the words IGNITABLE MIXTURE. Such a mixture is capable of spreading flame away from the source of ignition but it will not support continuous combustion until the temperature has reached the FIRE POINT. The FIRE POINT is generally a few degrees higher than the flash point. Since liquids can generate explosive vapors at their flash point, both points are important to the fire fighter.

The term flash point is applied most often to liquids but there are a few solids that produce flammable vapors, and they too have flash points. Camphor and naphthalene are two such solids.

TABLE 2–3. PROPERTIES OF SOME COMMON FLAMMABLE LIQUIDS

	FLASH POINT	IGNITION TEMP.	LOWER FLAMMABLE LIMIT (% BY VOL. IN AIR)	UPPER FLAMMABLE LIMIT (% BY VOL. IN AIR)	SPECIFIC GRAVITY (WATER=1.0)	VAPOR DENSITY (AIR=1.0)	BOILING POINT	WATER SOLUBLE
UL Class 90: Acetone (Dimethyl ketone) CH_3COCH_3	0° F. 18° C.	1000° F. 538° C.	2.6	12.8	0.8	2.0	134° F. 57° C.	Yes
Acetaldehyde (Ethanol or Acetic aldehyde) CH_3CHO	−36° F. −38° C.	365° F. 185° C.	4.1	55.0	0.8	1.5	70° F. 21° C.	Yes
UL Class 55-60: Amyl acetate—iso (banana oil) $CH_3COOCH_2CH_2CH(CH_3)_2$	77° F. 25° C.	715° F. 379° C.	1.0	7.5	0.9	4.5	290° F. 143° C.	Slightly
UL Class 110 plus: Carbon disulfide (Carbon bisulfide) CS_2	−22° F. −30° C.	212° F. 100° C.	1.3	44.0	1.3	2.6	115° F. 46° C.	No

TABLE 2–3. PROPERTIES OF SOME COMMON FLAMMABLE LIQUIDS (CONT.)

	FLASH POINT	IGNITION TEMP.	LOWER FLAMMABLE LIMIT (% BY VOL. IN AIR)	UPPER FLAMMABLE LIMIT (% BY VOL. IN AIR)	SPECIFIC GRAVITY (WATER=1.0)	VAPOR DENSITY (AIR=1.0)	BOILING POINT	WATER SOLUBLE
UL Class 70: Ethyl alcohol (Ethanol or Grain alcohol) C_2H_5	55° F.* 13°C.	793° F. 423°C.	4.3	19.0	0.8	1.6	173° F. 78°C.	Yes
Ethyl amine (Amino ethane)—70% aqueous solution—$C_2H_5NH_2$	<0° F. <−18°C.	723° F. 384°C.	3.5	14.0	0.8	1.6	62° F. 17°C.	Yes
UL Class 100: Ethyl ether (Diethyl ether or Ethyl oxide) $C_2H_5OC_2H_5$	−49° F. −45°C.	356° F. 180°C.	1.9	48.0	0.7	2.6	95° F. 35°C.	Slightly
Gasoline C_5H_{12} to C_9H_{20}	−36° to −45° F.** −38° to −43°C.	536° to 853° F. 280° to 456°C.	1.4	7.4 to 7.6	0.8	3 to 4	100° to 400°F. 38° to 204°C.	No

*As ethyl alcohol is mixed with water, the flash point changes. 50% alcohol and 50% water has a flash point of 75°F. (24°C.). 10% alcohol and 90% water has a flash point of 120°F. (49°C.).

**The properties of gasoline vary according to the octane.

Testing bureaus use two methods to determine flash points. In one, vapors from liquids are ignited in open cups (OC); in the other, in closed cups (CC). Open cup determinations—generally higher in temperature for the same liquid—more closely approach the flash point of a flammable liquid in the open air.

Flash point is used as the primary measure of flammability hazard by authorities who deal with the storage, use, or transportation of flammable liquids. The Department of Transportation (DOT) requires a *red label* on any flammable liquid with a flash point at or below 80°F. (27°C.) (OC). In some parts of the United States, daytime temperatures reach or exceed this figure at least a few times during the year. Red label flammable liquids, in such areas, are ready to ignite because their flash points have been reached.

The greatest importance of flash point to the fire fighter is that it indicates the proper method of extinguishment.

Use of Water on Flammable Liquids

For years, there has been strong disagreement about the use of water on low flash point liquids such as gasoline. Some have contended that water fog, properly used, will extinguish a burning gasoline spill. Others say that fog cannot do the job. They say that reported extinguishment occurred because the fuel burned out completely, rather than because of the application of water. Enough foam or dry chemical, applied correctly, can extinguish a gasoline fire, but how much can be done with water? Can you knock out a gasoline fire with water fog? In considering these questions, we must keep in mind that a layer of gasoline one inch thick will burn itself out in about five minutes; a foot of gasoline, or eight inches of kerosene, will burn out in one hour.

Reported cases of fires in tank trucks that were carrying thousands of gallons of gasoline seem to support those who argue in favor of water. For example, a Florida fire in a transport truck-trailer, carrying 5400 gallons of gasoline, began when a tire collapsed and ignited. Fire spread to the discharge-and-pump compartment. The accident occurred 15 miles from a fire station and 34 minutes passed before the alarm came in.

Arriving at the scene, fire fighters used two 1½-inch lines with nozzles set for a 60-degree fog pattern; discharge was 75 gpm with a nozzle pressure of about 125 pounds. Hitting the fire beneath the truck cooled the tank, and, after 24 minutes of this application, the fog was directed to the vents on top of the truck. Within five minutes, the fire was completely extinguished.

In spite of incidents like this, many fire fighters remain unconvinced that water is effective on fires involving gasoline and other low flash point liquids because other fire cases tend to contradict the theory that water can extinguish a gasoline fire.

In one California city, a gasoline spill ignited in the middle of a busy street. The spill was attacked with two 1½-inch fog lines. The burning gasoline was pushed across the street into a store which ignited. Realizing their mistake, the fire fighters worked their lines into the front of the store and tried to push the gasoline back into the street. Floating on the water, the burning gasoline ran across the street into another store, setting it on fire. Several minutes later, the gasoline burned itself out.

Keep a few temperatures in mind when entering a firehouse debate on the use of water on a gasoline fire: gasoline's flash point is around −45°F. (−43°C.) while the temperature of hydrant water is approximately 60°F. (16°C.) The boiling point of gasoline varies between 100°F. (38°C.) and 400°F. (204°C.). These temperatures make it plain that gasoline gives off flammable vapors at low temperatures. It will produce these vapors at temperatures when water would have long since turned to ice. It is these vapors that burn.

Let us now suppose that these vapors become ignited. Although the heat of burning gasoline vapor is high, the surface of the liquid itself cannot rise much above its boiling point, which is an average of 250°F. (121°C.), or else all the liquid would turn to gas. (Water boiling in an open pot cannot rise above 212°F. (100°C.) no matter how much heat is applied; it simply boils away more rapidly.) Although water at 60°F. (16°C.) can lower the temperature of the liquid gasoline to below its boiling point, this is unimportant. The vapors which concern us are being produced at the flash point of gasoline, whereas the temperature of hydrant water is 100 degrees Fahrenheit (38 degrees Celsius) higher than this and, therefore, totally incapable of stopping vapor production. Further, since it is steam formation that makes water most useful in attacking and cooling the seat of a fire, a gasoline fire presents a real problem. The flammable liquid itself is generally not hot enough to form much steam. Since water is heavier than gasoline it will sink beneath the surface, accomplishing nothing but agitation and possible overflow. Most steam production occurs where it will do little good, in the burning vapor mixtures that quickly rise out of the combustion zone.

Worse yet, water fog directed into the burning vapor may temporarily *increase* intensity of burning. Vapor just above the gasoline may be too rich to burn (accounting for fire in the smoke well away from the liquid), but water fog may add enough oxygen to bring the vapor down to the burnable level. And, if steam does momentarily extinguish part of the burning gasoline vapors, the effect is usually short. The heat of a gasoline fire rapidly raises metal and other exposed materials to temperatures that cause immediate reignition. Although water can be extremely useful in cooling these exposures, eventually reducing the amount of heat being produced, whether it will extinguish a gasoline fire is highly problematic.

As the NFPA says: "The ability of water to effect extinguishment is limited. Any water reaching the surface of a low flash point liquid that is burning will probably not turn to steam. If the burning liquid is in a tank, the water may cause the tank to overflow; in the case of a spill fire, the water will probably cause the fire to spread."

Summarizing, the use of water on low flash point liquids is a questionable procedure. But there would be no debate if water had never been known to extinguish a gasoline fire. When used by competent fire fighters, under ideal conditions, fog has been known to displace the flames on the surface of flammable liquid spill, just like blowing out a candle. There are also those who feel it is possible to supersaturate the air with moisture above a spill or tank fire, so that the gasoline is, in effect, under water, where it cannot burn. They envision a series of master fog streams, covering the entire surface of the burning liquid, displacing the air with an atmosphere of water vapor or steam.

These veteran fire fighters agree that water cannot extinguish a low flash point flammable liquid *by cooling*. However, after attending Navy fire-fighting school and oil fire schools, they have witnessed the extinguishment of gasoline tank fires by smothering with steam. Two things are necessary: the tank requires outage—space between the liquid level and the top of the tank to accumulate steam—and the fire must burn long enough to heat the surface of the flammable liquid and tank surfaces. They also point to some of the other uses of water: to cool the surface of a flammable liquid before foam is applied, to direct the flow of flammable liquids to desirable areas, and to extinguish ground fires and cool containers, as in the truck tank fire mentioned above, so that vapor formation will be lessened.

Most authorities recommend the use of the proper types of foam or dry chemical on low flash point flammable liquids, with a possible supporting water-fog pattern to cool exposed surfaces.

Water is much more effective on flammable liquids with a flash point between 100°F. and 212°F (38°C. and 100°C.), well above normal hydrant water temperature. In these cases, water is capable of cooling the liquid to below its flash point, of stopping vapor production, and of putting out the fire.

On combustible liquids with a flash point above 212°F. (100°C), care must be taken not to allow water to become trapped below the surface of the burning liquid. Because of the flash point, we know that the liquid itself is already above the boiling point of water. If straight streams of water are injected below the surface of the burning liquid, deliberately or by accident, the water will quickly turn to steam, increasing 1700 times in volume. A steam explosion is possible (for this is not unlike applying water casually to red-hot metal), as is an overflow or "slopover," as the expansion of the water forces the flammable or combustible liquid over the edges of the tank.

On liquids with a high flash point, excellent results have often been obtained by the careful use of fog patterns that cool the liquid without allowing water to become trapped beneath the surface.

Whatever your opinion on the use of water on a gasoline fire, all this discussion stems from a single characteristic of flammable liquids: *flash point*.

Remember: Flash points can vary, depending on factors such as the pressure above the liquid or the oxygen content of the atmosphere. Flash point values are calculated on the basis of normal atmospheric pressure at sea level, 14.7 pounds per square inch (760 torrs). If the pressure is less than one atmosphere, or if the supply of oxygen is increased for any reason, the flash point may be lowered considerably.

Remember: Flash point is related to flammable limits and vapor pressures. These relationships will be discussed later in the chapter.

Ignition Temperature

Ignition temperatures and flash points are entirely different, although they are often mistaken for each other. As we have seen, all flammable liquids have a flash point, the temperature to which they must be heated before they give off enough vapor to be ignited. Flammable gases have no flash point and require no preheating of any kind, only an *ignition source*. Flammable gases may be thought of as liquids that reach their flash points and boiling points at very low temperatures. Flammable solids, with few exceptions, have no flash points either.

But, all flammable liquids, solids, and gases have an *ignition temperature*, which the NFPA defines as "the minimum temperature required to initiate or cause self-sustained combustion independently of the heating or heated element." (In other words, anything flammable will burn at a certain temperature without an additional ignition source.) Flash point only tells us when sufficient vapors will be present to burn. They still require a source of ignition.

Ignition Source vs. Ignition Temperature Any ignition source must be at least as hot as the ignition temperature of the material. A flame must be hot enough to heat vapors to their ignition temperature in the presence of air; sparks must be intense enough and last long enough to ignite flammable vapor-air mixtures; hot surfaces must be large enough and hot enough, and the liquid must remain in contact with the hot surface for a sufficient length of time.

In practice, we find most common ignition sources have a temperature much higher than the ignition temperature of most flammable substances. A match flame is over 2000°F. (1093°C.); electric arcs are about the same temperature. A cigarette ranges from 550°F. to 1350°F. (288°C. to 732°C.).

In theory, every flammable liquid has an ignition temperature. (Some explode before this temperature is reached.) The *ignition temperature of a liquid has no relationship to its flash point.* As Table 2–3 shows, there is a difference of 1000 degrees Fahrenheit (538 degrees Celsius) between acetone's flash point and ignition temperature. By contrast, for carbon disulfide the difference is less than 250 degrees Fahrenheit (121 degrees Celsius). Carbon disulfide has such an extremely low ignition temperature, 212°F. (100°C.), that its vapors can be ignited by sources generally considered safe. Since a steam pipe or an electric light bulb will reach temperatures above 212°F. (100° C.), special storage requirements are needed for carbon disulfide.

Ignition Temperatures of Flammable Gases Acetylene ignites at 571°F. (299°C.), methane at 999°F. (537°C.). What is surprising is that many flammable liquids have *lower* ignition temperatures than certain flammable gases. Methane's ignition temperature, for example, is much higher than that of gasoline.

Ignition Temperatures of Flammable Solids Most woods ignite around 400°F. (204°C.) especially if held at that temperature for any time. Paper, cotton, woolen blankets, rayon, and fiberboard ignite somewhere between 400°F. and 500°F. (204°C. and 260°C.).

Ignition Temperatures Vary Ignition temperatures, like flash points, change quickly with conditions. The amount of oxygen in the air, where and how the ignition occurs, the duration of heating, the speed of the heat rise, and other factors can all affect the ignition temperature of a material. *Treat all figures quoted in various sources as approximations.* The ignition temperature of one flammable liquid, as determined by three different methods, varies by more than 500 degrees Fahrenheit (260 degrees Celsius).

Remember: The ignition temperature of a material is the temperature at which it will ignite without an additional ignition source.

Remember: The ignition temperature of a material is the temperature an ignition source must have in order to ignite a material. (This is really saying the same thing backwards.)

Flammable Range

Situations often occur in which flammable vapors from a low flash point liquid and an ignition source are both present, yet no fire or explosion results. Some typical situations of this sort are the lighting of a cigarette by an attendant or customer while the tank is being filled in a service station and the cleaning of an article near the pilot light of the hot water heater in a home.

Since a vapor will explode if it can, there is clearly some condition present that prevents ignition. This saving factor is often the *explosive limit* of a flammable gas or vapor. In terms of our automobiles, a mixture of vapor and air below the lower flammable limit is too lean to burn; a mixture of vapor and air above the upper flammable limit is too rich to burn.*

Table 2–3 shows that concentrations of gasoline vapor in air below 1.4 percent are too lean to burn. Concentrations of vapor above 7.6 percent are too rich to burn. All concentrations between these two figures are flammable. This spread between lower and upper flammable limits (6.2 percent for gasoline) is known as the *flammable* or *explosive range*. Once again, there is wide variation among the various liquids and gases. Gasoline has a relatively narrow flammable or explosive range. Every other common flammable liquid listed in Table 2–3 has a wider flammable range. Some flammable gases exceed the ranges listed. Acetylene, for example, has a lower flammable limit of 2.5 percent and an upper flammable limit of 81 percent, a range of 78.5 percent.

Factors Affecting the Flammable Range Increases in temperature increase the flammable range. Increases in pressure above one atmosphere may or may not affect the flammable range, but substantial decreases in pressure almost always narrow it. The amount of available oxygen greatly affects the flammable range. Flammable anesthetic gases, when mixed with oxygen or nitrous oxide, have a greatly increased explosive range. Certain gases and vapors may form flammable mixtures in atmospheres other than air or oxygen—in chlorine, for example.

Flammable Range and Flash Point There is a direct relationship between flammable limits and flash points. The flash point of a liquid is the temperature at which there is enough vapor production (evaporation) to reach the *lower* flammable limit. Evaporation increases with temperature. The amount of vapor in the air passes completely through the flammable range until the upper flammable limit is reached and exceeded at higher temperatures.

> An understanding of the flammable limits and flammable range will enable the fireman to do a more effective job in fire prevention, and also will enable him to more effectively understand and fight flammable liquid fires. For example, at average temperature and pressure, the vapor-air mixture above low flash point liquids, such as gasoline in closed tanks and containers, is usually too "rich" to burn. For liquids whose flash points are above the storage temperature, the vapor-air mixtures are too "lean" to be ignited. For liquids whose

---

*Flammable limit figures are based upon normal atmospheric temperatures and pressures.

> flash points approximate the storage temperatures, the vapor-air mixtures are likely to be within the flammable limits. Examples of these three conditions for common liquids which are stored in tanks and containers are: gasoline (too rich); ethyl alcohol or methyl alcohol (within the flammable limits); and kerosene (too lean).

Notice the emphasis on *closed containers* in this quotation from the "Flammable Liquid Notebook," *Fireman's Magazine,* August, 1955.

Open Spills When a flammable liquid spill occurs in the open, however, the entire flammable range is present. A single gallon of gasoline will raise 2300 cubic feet of air to the lower flammable limits. But the mixture is uneven. Directly above the spill, the vapor-air mixture may be too rich. Farther away, the mixture is within the flammable range, finally becoming too lean at still greater distances. Flames occur in the region where vapors are in the flammable range. This may be some distance from the liquid. As we saw in our discussion of flash point, burning rate is often increased when water fog is first applied to a low flash liquid fire. This is because an inrush of air from the nozzle suddenly lowers a rich mixture into the flammable range.

Do not rely on your nose to tell you if the lower limits of flammability have been reached. There may not be a noticeable odor because many liquids and gases are practically odorless. Methane, for example, has to have an odorant added so there will be a warning of a natural gas leak that will aid in its detection.

Physical Properties and Extinguishment

Water is the fire fighter's first line of defense. When we approach an emergency, our training and equipment condition us to think of using water first. Only if we are convinced that the use of water is inadvisable will we turn to other extinguishing agents. Anything, therefore, that affects the usefulness of water should concern us. One such factor is whether a particular gas or liquid will mix with water. Some liquids, such as acetone, ethyl alcohol, and methyl alcohol, are totally soluble (miscible) in water. This solubility is infinite, meaning that any proportion of these liquids and water will mix. Nonmiscible flammable liquids present a greater hazard.

The Dilution Technique

Fire in a water-soluble liquid can be fought by entirely different methods than can fire in a liquid that is not water soluble. One such method is by *dilution,* adding water to the liquid. Looking at Table 2–3 under ethyl alcohol (footnote), we can see that as the percentage of water goes up in an alcohol-water mixture, so does the flash point. Fifty percent alcohol in water has a flash point of 75°F. (24°C.); ten percent alcohol in water has a flash point of 120°F. (49°C.).

In a spill fire, this dilution technique can be used to raise the flash point of a burning liquid so high that it will no longer produce flammable vapors. If the water-soluble liquid is not on fire, application of water may lower the temperature of the mixture below the level sufficient to produce potentially explosive vapors. This is another place where the high specific heat of water is of value to the fire fighter.

The only problem in using the dilution technique involves the amount of water required. This varies from liquid to liquid. Ethyl alcohol requires the addition of from five to six times its volume of water to raise the flash point above 100°F. (38°C.). To dilute a 55-gallon drum of spilled alcohol, we would need at least the contents of a 300-gallon booster tank. This amount of fluid could cause difficulties, and many liquids require even more dilution. However, in a situation where there are adequate supplies of water and no drainage problems, dilution is a fairly quick way to greatly reduce the hazards of spills. Dilution is of less value in a flammable liquid tank fire. The required volume of water will quickly overflow the tank in most cases.

Using Foam

If foam is chosen to fight a fire in a water-soluble liquid such as alcohol or acetone, it must be of a special type. Regular foam breaks down rapidly on contact with these liquids; a special "alcohol" foam should be used instead. Alcohol foams can be generated from powders or liquid concentrates. While they are designed to be less liable to breakdown, they generally require a gentle application to the surface of the water-soluble liquid. Further, alcohol foams are generally incompatible with regular foam, and, if the two are mixed, they can gum up equipment badly. "Wet water" should not be used on flammable or combustible liquid fires where the liquids are water soluble. Surprisingly, industrial concerns are often unaware of the need for foam of a particular type. They may store regular foam when most of the solvents they have on hand are water miscible. Consider this possibility during inspections.

Aqueous Film-forming Foaming Agents, known to fire fighters as AFFF, are foam products that are especially valuable in combating liquid hydrocarbon fuel fires. AFFF develops a low viscosity, fast-spreading surface barrier that provides a floating film beneath the foam. This aqueous layer helps to suppress the combustible fuel vapors and to cool the fuel substrate.

AFFF was developed by the 3M Company of St. Paul, Minnesota, in conjunction with the U.S. Navy and the Air Force to quell blazes on ships and aircraft. The Los Angeles City Fire Department was the first to put

AFFF to use in nonmilitary fire fighting. This was in 1972 at the GATX loading dock fire in San Pedro (see Figure 2–1) where a number of 30- to 80-thousand-gallon tanks of butyl acetate, ethyl acetate, methyl ethyl ketone, and other highly flammable liquids were involved in fire. A little more than half an hour after the L.A.F.D's Light Water unit arrived, the fire was out. (See Chapter 3 for further details.)

FIGURE 2–1. Battling Gasoline Loading Dock Blaze (Photo by Dale Magee)

Fog Patterns

Water-soluble vapors and gases can sometimes be swept from the air with fog patterns. This offers an excellent way to reduce the explosive and toxic potentiality of the contaminated atmosphere around a fire. (Spray patterns can also be used to dissipate nonsoluble gases or vapors by the force of the stream.)

Some liquids are only partially soluble in water. Many acetates and alcohols will only mix in small and variable amounts. The problem of what type of foam to use on "very slightly soluble" liquids is usually left unsolved. Most authorities simply recommend trying another type of foam if one does not work.

Remember: Many liquids, such as gasoline, kerosene, and oil, are not soluble in water. Flammable liquids of this type cannot be extinguished by dilution. Regular foam is designed for use on these liquids.

Specific Gravity

As we use the term, SPECIFIC GRAVITY means the weight of a solid or liquid substance, as compared to the weight of an equal volume of water. The specific gravity of water is therefore one (1.0). A liquid or solid with a specific gravity of less than one will float on water; if its specific gravity is more than one, it will sink. Water solubility of a liquid has some bearing on whether it will sink or float.

Most flammable liquids (but not all) are lighter than water. Again, gasoline is an example. Burning gasoline, floating on top of running water, can spread fire at an awesome rate. Several reports of major flammable liquid fires speak of "running rivers of fire" pouring from a building. The danger to exposures is obvious.

Blanketing

Knowledge of the specific gravity of a liquid can be a useful fire-fighting tool. If there is a leak in the bottom of a tank containing a liquid of low specific gravity, a hose stream into the tank can float the flammable liquid above the level of the leak. If a liquid is heavier than water and not soluble in water, it is sometimes possible to float a layer of water on top of the burning liquid. This method is called BLANKETING. The water is generally applied with an open-hose butt, under low pressure, for gentle application. Fires in carbon disulfide (specific gravity: 1.3) have been extinguished by this method.

Vapor Density

Vapor density is the relative density of a vapor or gas (with no air present) compared to air. Air is rated as one (1.0). A figure of less than one indicates a vapor or gas is lighter than air. A figure greater than one indicates a vapor or gas is heavier than air. Vapor density figures can be misleading because they are laboratory comparisons; there is no air or vapor movement present during these measurements. In the field, we seldom find a pure gas or vapor; we deal with mixtures of vapor and air. These mixtures will not have the same vapor density as pure vapors or gases. Understandably, their weight will be closer to that of air. Further, vapors and gases become lighter when heated by fire, and hot, upwelling air currents can lift heavy gases or vapors.

Remember: Vapor density figures are not necessarily an indication of the behavior of the vapors.

Ordinarily, flammable liquids with a specific gravity that is greater than water (greater than 1.0) are less hazardous, since they sink in water and would have less tendency to spread their vapors.

Vapor Spread

In spite of these qualifications, a knowledge of vapor density can be useful to the fire fighter. The density of an escaping vapor or gas will determine where it will spread. It may enable us to predict where gas will collect—near the ceiling or in a basement. It may also show the probable future course of an explosion or fire and where to look for potential ignition sources. For example, ethyl ether has a vapor density of 2.6, making it much heavier than air. It will collect near the floor if undisturbed. For this reason—to guard against ether explosions—electrical outlets in hospital operating rooms are high up on the walls.

Some gases are lighter than air and some are heavier than air. Lighter-than-air gases and vapors tend to spread farther but also disperse much more rapidly. This dispersion is called the RATE OF DIFFUSION. Almost without exception, however, vapors from flammable liquids are heavier than air and tend to stick together in a cloud. (Other heavier-than-air gases act the same way.) AIA Special Interest Bulletin 157 outlines the dangers of this type of vapor cloud.

> Investigations of fires involving flammable liquids have frequently indicated the cause as the ignition of the vapor at varying distances from the sources. . . .
>
> The possibility of any given vapor igniting at distances in the open air will largely depend upon the properties of the vapor and atmospheric conditions existing at the time. . . . When the vapor is heavier than air, its rate of diffusion will be slow, but if the vapor is equal to or lighter than air, diffusion will be rapid. . . . This means that heavy vapors or gases . . . under certain atmospheric conditions are capable of traveling considerable distances in the open before igniting. . . .
>
> Atmospheric conditions tending to promote such conditions are . . . weather extremes . . . a hot, humid summer day with little or no air movement, and days of still, extreme cold.

A perfect example of this hazard occurred in a 1915 fire: A tank car, loaded with casing-head gasoline, stood over a weekend in the Santa Fe Freight yards at Ardmore, Oklahoma. Monday was hot and humid, with very little air movement. Heated by the sun, the gasoline expanded and vaporized, creating enough pressure to force the tank's safety valve open and allow the gasoline vapor to escape into the air. This condition continued for several hours, the vapor spreading and filling many buildings with a combustible air-vapor mixture.

At some unknown location in the area, ignition occurred. The subsequent explosion wrecked all buildings within 400 feet (122 meters) of the tank car, cracked many walls, and destroyed parapets and chimneys on buildings 1,200 feet (366 meters) away. The exact source of ignition was never determined but sources of flame included a switch light 215 feet (65 meters) from the car; an asphalt plant in operation 300 feet (91 meters) to the south; a locomotive switching in the yards to the north.

People described the flame as "fire bounding along the ground in streaks," igniting women's clothing as far as 350 feet (107 meters) away from the tank car, filling the streets 1,000 to 1,200 feet (305 to 366 meters) away; penetrating the surrounding buildings and towering as high as a skyscraper in the air.

After reading this account, there can be little doubt that vapor clouds can and do travel and that ignition can take place at a point far from the vapor source. Records bear this out; the Association of American Railroads' Bureau of Explosives indicates that 81 fires, from 1919 to mid-1929, were caused by vapor travel and subsequent ignition.

Fire chiefs and others having to do with fire prevention must recognize this danger. Many fire department officers and inspectors overlook the element of topography in passing on plans for storage and handling operations. Venting and distance from buildings are not enough. Vapors of many petroleum products, being heavier than air, tend to settle to the ground where they spread out and run toward low points.

Boiling Point

In this discussion of boiling point, we have to deal with three concepts: BOILING POINT itself, VAPOR PRESSURE, and EVAPORATION RATE. All three are interrelated and not really difficult to understand.

Going back, in the section on heat (see pages 9 and 10), we discovered that the difference between a liquid and a gas is the difference in the vibratory rate of the molecules. As temperature rises, the vibratory rate increases, and at the *boiling point,* molecules have absorbed so much energy that they stop sliding over each other and fly off in all directions. Now, we are going to add a refinement to our definition of temperature: Temperature is the measure of the *average* vibratory rate in a group of molecules.

In any liquid, some molecules are moving faster than the average, some slower. At any temperature, a certain number of the more rapidly vibrating molecules escape from the liquid and turn into vapor. If the liquid is in the open, these vapor particles move away from the liquid. This is called evaporation. All liquids evaporate but at different rates. Evaporation rate is the rate of change from a liquid to a vapor at temperatures below the boiling point. The lower the boiling point, the more volatile and therefore the more hazardous the flammable liquid.

Vapor Pressure

If a liquid is confined, however, escaping molecules of vapor build up pressure in the limited space above the liquid. At first, there are only a few vapor molecules in this space. Then, as more and more molecules turn into vapor, their random movement causes an increasing number to re-enter the liquid. Eventually, the number of arriving and departing molecules reaches an equilibrium. The pressure exerted by the escaping vapor against the sides of the container at this point is called the *vapor pressure* of the liquid. Generally, it is measured in pounds per square inch. (Psig stands for pounds per square inch gage. Gage pressure does not include the normal atmospheric pressure of 14.7 pounds. If atmospheric pressure has been included it is called absolute pressure and abbreviated psia.) Gas pressure or vapor pressure is nothing more than the impact of molecules against some restraining vessel. The more molecules, the more impacts and, hence, the more pressure.

Vapor pressure varies with temperature. The higher the temperature, the greater the number of molecules that move fast enough to escape, because the average vibratory rate continues to rise. Higher vapor pressure also reflects the fact that faster-moving molecules cause more impacts and more pressure.

The rate at which a liquid actually turns to vapor also depends upon the opposing pressure. Generally, this opposing pressure is supplied by the earth's atmosphere, 14.7 pounds per square inch at sea level. When the temperature of a liquid is increased so that the vapor pressure equals the atmospheric pressure, the material reaches its boiling point. Now, all of its molecules are vibrating fast enough to begin flying. A continuous stream of vapor bubbles occurs in the liquid when it is heated in an open container. As the opposing pressure is lower at high altitude, liquids at high altitudes boil at lower temperatures.

Boiling point is an indication of a material's readiness to turn into vapor (volatility). The amount of vapor produced by a liquid at any temperature is directly related to its vapor pressure and its boiling point. In general, the lower the boiling point of a liquid, the higher its vapor pressure and evaporation rate. In liquids with similar flash points, the one with the lower boiling point is considered more hazardous. This is reflected in the NFPA Class ratings given to flammable liquids (see Table 2–1).

Knowledge of the vapor pressure of a particular liquid can be of considerable help to the fire fighter. First, it allows an estimate of how much vapor is present above the liquid in a closed container. Second, by dividing vapor pressure by 14.7, one can determine the percentage of vapor present.

Whether this percentage is within the flammable range or not is easily determined from standard tables. The higher the vapor pressure, the higher the rate of evaporation and the more hazardous the flammable liquid.

Sometimes, a nonflammable liquid is added to a flammable liquid to raise its flash point above a dangerous level. But, if the vapor pressure of the additive is higher, this may only give temporary protection. The additive will evaporate more rapidly, and, over a period of time, the flash point will continue to lower until the flash point of the final mixture approaches that of the original.

Finally, there is no sharp line between a gas and a liquid. We use an arbitrary division based upon vapor pressure. A liquid is defined as a fluid having a vapor pressure not exceeding 40 pounds per square inch (2068 torrs) at 100°F. (38°C.).

Summary

At the beginning of this chapter, we asked the question, Why is a liquid flammable? We now see that the characteristics (properties) of each liquid determine how and why it will burn. If the temperature is high enough, every flammable liquid will produce vapors within the flammable range. If an ignition source of proper intensity is present, the vapors will ignite or explode unless some other condition prevents this at the moment. Remember that conditions can change almost without warning and very rapidly.

Methods used to combat the hazards of flammable liquids must be based upon the answers to such questions as:

- What is the flash point of the liquid?
- Is this liquid or vapor soluble in water?
- What is this liquid's specific gravity?
- What is the vapor density of this vapor or gas?
- Where are ignition sources located?

QUESTIONS ON CHAPTER 2

What is the importance of each of the following properties of flammable liquids:

a. Flash point
b. Ignition temperature
c. Flammable range
d. Water solubility
e. Specific gravity
f. Vapor density
g. Boiling point
h. Vapor pressure

RECOMMENDED READING

NFPA/Bulletin 325, "The Fire Hazard Properties of Flammable Liquids, Gases and Solids."

VISUAL AIDS

"Chemistry of Fire Fighting," produced by U.S. Navy during World War II. 45 minutes, black and white. Demonstrates many properties of flammable liquids.

DEMONSTRATIONS

Four important properties can be demonstrated, using four easily obtained flammable liquids: gasoline, kerosene, carbon disulfide, and acetone.

1. *Flash point* can be shown by holding a match over Pyrex beakers of gasoline and kerosene. When the gasoline ignites, the kerosene can be heated by the gasoline flame. Soon the kerosene can be ignited also, demonstrating the danger of exposure fires.
2. *Ignition temperature* can be shown with gasoline and carbon disulfide. The heat of a burning cigarette probably will not ignite gasoline vapor. But it is possible for ignition to occur. Therefore, the cigarette should be held in a gloved hand. The cigarette will immediately ignite the carbon disulfide.
3. *Specific gravity* can be demonstrated by the difference between gasoline and carbon disulfide. If water is added to gasoline, the gasoline will float and can be ignited. If water is carefully applied to carbon disulfide and allowed to stand for a few minutes, the carbon disulfide will sink and ignition will not be possible.
4. *Water solubility* can be shown by the gasoline already used and by acetone. Stir the beaker containing gasoline and water and they will immediately separate. Ignite a small quantity of acetone in a beaker and extinguish it by covering with an asbestos pad. Add enough water to dilute the acetone well and it will not re-ignite, for dilution raises its flash point.

PRECAUTIONS

Practice these demonstrations in an open area before class begins. Have several fairly large Pyrex beakers to contain the liquids and vapors and asbestos pads for extinguishment. A small dry chemical extinguisher should be kept handy to take care of any mishaps. Remember that carbon disulfide, in particular, is toxic and has an extremely low ignition temperature. Do not allow quantities of its vapor to escape.

3 Flammable Liquids

In Chapter 2, we discussed many of the properties shared by all flammable liquids. Table 2–3 lists eight common flammable liquids, but each of these represents a much larger class. Acetone is an example of a KETONE. Isoamyl acetate is an ESTER. Also included are an ETHER, an ALCOHOL, an AMINE, an ALDEHYDE, and a HYDROCARBON. In Chapter 3, we will consider how these classes of liquids differ from each other in uses and hazards.

Hydrocarbons

Crude oil pumped from the ground is a mixture of many different molecules. It is refined into usable products (called FRACTIONS by oilmen) by an industrial process called *fractional distillation*. The various molecules are separated, collected at different levels of a fractionating tower, piped away, cooled, and blended into familiar products. (See Table 3–1.)

Gasoline, kerosene, naphtha, petroleum ether, fuel oil, lubricating oils, asphalt, and other liquid petroleum fractions, the subject of this section, are all hydrocarbons. Their molecules are made up of hydrogen and carbon.

Let us first consider the remarkable carbon atom. Carbon atoms can link together to form long chains, containing hundreds or even thousands of atoms. We are concerned here with chains that are much shorter.

Paraffin Series

The difference between methane and ethane is that ethane has one more carbon atom in the chain, along with two more hydrogen atoms. This is also

TABLE 3–1. PROPERTIES OF TYPICAL PETROLEUM FRACTIONS

	FLASH POINT	IGNITION TEMP.	FLAMMABLE LIMITS (PERCENT BY VOLUME IN AIR)		SPECIFIC GRAVITY (WATER=1.0)	VAPOR DENSITY (AIR=1.0)	BOILING POINT
			LOWER	UPPER			
Asphalt typical (petroleum pitch)	400° F.* 204°C.	905° F. 485°C.			1.0–1.1	—	700° F. 371°C.
Crude petroleum	20° to 90° F. −7° to 32°C.				1.0		
Fuel oil #1 (Kerosene) (coal oil: range oil)	100° F.** 38°C.	444° F. 229°C.	0.7	5.0	1.0 (or less)	—	304° to 574° F. 151° to 301°C.
Fuel oil #2	100° F.** 38°C.	494° F. 257°C.			1.0 (or less)	—	
Fuel oil #4	130° F.** 54°C.	505° F. 263°C.			1.0 (or less)	—	
Fuel oil #5	130° F.** 54°C.				1.0 (or less)	—	
Fuel oil #6	150° F.** 66°C.	765° F. 407°C.			1.0 (more or less)	—	
Gasoline (60 octane)	−45° F. −43°C.	536° F. 280°C.	1.4	7.6	0.8	3.0–4.0	100° to 400° F. 38° to 204°C.
(100 octane)	−36° F. −38°C.	853° F. 456°C.					

TABLE 3–1. PROPERTIES OF TYPICAL PETROLEUM FRACTIONS (CONT.)

	FLASH POINT	IGNITION TEMP.	FLAMMABLE LIMITS (PERCENT BY VOLUME IN AIR)		SPECIFIC GRAVITY (WATER=1.0)	VAPOR DENSITY (AIR=1.0)	BOILING POINT
			LOWER	UPPER			
Lubricating oil (motor oil)	300° to 450° F. 149° to 232°C.	500° to 700° F. 260° to 371°C.			1.0 (or less)	—	680° F. 360°C.
Mineral oil	380° F. 193°C.				0.8–0.9	—	680° F. 360°C.
Naphtha safety solvent (Stoddard solvent)	100° to 140° F. 38° to 60°C.	300° to 450° F. 149° to 232°C.	1.1	6.0	0.8	—	357° F. (or higher) 181°C. (or higher)
Naphtha V.M.&P. (50 flash)	50° F.*** 10°C.	450° F.*** 232°C.	0.9	6.7	1.0	4.1	240° to 290° F. 116° to 143°C.
(High flash)	85° F.*** 30°C.	450° F.*** 232°C.	1.0	6.0	1.0	4.3	280° to 350° F. 138° to 177°C.
(Regular)	28° F.*** −2°C.	450° F.*** 232°C.	0.9	6.0	1.0	—	212° to 320° F. 100° to 160°C.
Petroleum ether (Petroleum naphtha, Benzine, Ligroin)	−50° to 0° F. −46° to −18°C.	550° F. 228°C.	1.1	5.9	0.6	2.5	95° to 140° F. 35° to 60°C.

TABLE 3–1. PROPERTIES OF TYPICAL PETROLEUM FRACTIONS (CONT.)

	FLASH POINT	IGNITION TEMP.	FLAMMABLE LIMITS (PERCENT BY VOLUME IN AIR)		SPECIFIC GRAVITY (WATER=1.0)	VAPOR DENSITY (AIR=1.0)	BOILING POINT
			LOWER	UPPER			
THE RING HYDROCARBONS							
Benzene (Benzol)	12° F. −11°C.	1044° F. 562°C.	1.3	7.1	0.9	2.8	176° F. 80°C.
Naphtha (coal tar)****	100° to 110° F. 38° to 44°C.	531°F. 277°C.			0.8	—	300° to 400° F. 149° to 204°C.
Naphthalene (white tar)	174° F. 79°C.	979° F. 526°C.	0.9	5.9	1.1	4.4	424° F. 218°C.
Toluene (Toluol)	40° F. 4°C.	997° F. 536°C.	1.2	7.1	0.9	3.1	231° F. 111°C.
Xylene, ortho	63° F. 17°C.	867° F. 464°C.	1.0	6.0	0.9	3.7	292° F. 144°C.
meta	77° F. 25°C.	982° F. 528°C.	1.1	7.0	0.9	3.7	282° F. 139°C.
para	77° F. 25°C.	984° F. 529°C.	1.1	7.0	0.9	3.7	281° F. 138°C.

* Depending on the curing qualities, the flash point of asphalt ranges from 100°F. to 250°F. (38°C. to 121°C.). Cutback asphalt has a flash point of 50°F. (10°C.).

** Minimum or legal flash point, which varies in different states.

*** Flash point and ignition temperature will vary depending on the manufacturer.

**** Coal tar naphtha is composed mainly of toluene and xylene.

```
    H           H   H           H   H   H           H   H   H   H
    |           |   |           |   |   |           |   |   |   |
  H-C-H       H-C - C-H       H-C - C - C-H       H-C - C - C - C-H
    |           |   |           |   |   |           |   |   |   |
    H           H   H           H   H   H           H   H   H   H
```

Methane CH_4 | **Ethane** C_2H_6 | **Propane** C_3H_8 | **Butane** C_4H_{10}

FIGURE 3–1. Simple Hydrocarbons

the difference between ethane and propane, or between propane and butane. (See Figure 3–1.) This type of carbon chain characterizes the PARAFFIN SERIES.

Each molecule in the paraffin series contains twice as many hydrogen atoms, plus two, as carbon atoms. This structure is an example of a SATURATED HYDROCARBON: all of the available carbon links are taken by hydrogen atoms. There are other hydrocarbon series that are UNSATURATED. We will discuss these—the OLEFIN SERIES and the ACETYLENE SERIES—later. Table 3–2 lists the members of the paraffin series up to a ten-carbon chain.

As the molecules in a particular series become larger, they naturally become heavier. Methane and ethane are gases. Propane and butane are gases which can be liquefied easily by pressure, the LP (liquid petroleum) gases. Butane, a gas at room temperatures, will liquefy on a cold day. When the chain becomes five carbon atoms long, the molecules are too large to be gases at ordinary temperatures. Pentane is a liquid, as are hexane, heptane, octane, nonane, and decane.* Eventually the carbon chain becomes so long that semisolids and solids are formed. Petroleum jelly (Vaseline), paraffin, and asphalt are examples.

All of the liquids in Table 3–2 are flammable and potentially explosive. The following incident shows their dangerous nature. In 1956, a half-million-gallon spheroid tank on a Texas tank farm was being filled with pentane and hexane. Some of the liquid spilled. An open fire heater, approximately 400 feet (122 meters) from the tank, ignited the vapors. There was a flashback along the vapor trail. For an hour, the spheroid tank was involved in a ground fire caused by the spilled liquid. Two fire departments responded. They used water on exposed tanks and extinguished fires in a nearby floating-roof gasoline tank several times. Water was also used to cool the involved spheroid tank.

---

*The names of this part of the paraffin series use Greek numbers to show the number of carbon atoms in the molecule. These Greek numerical prefixes are widely used in chemistry—*mono*, one; *di*, two; *tri*, three; *tetra*, four; *penta*, five; *hexa*, six; *hepta*, seven; *octa*, eight; *nona*, nine; *deca*, ten. *-ane* is the suffix for a saturated hydrocarbon.

TABLE 3–2. PARAFFIN SERIES

	FLASH POINT	IGNITION TEMP.	LOWER FLAMMABLE LIMIT (%BY VOL. IN AIR)	UPPER FLAMMABLE LIMIT (% BY VOL. IN AIR)	SPECIFIC GRAVITY (WATER=1.0)	VAPOR DENSITY (AIR=1.0)	BOILING POINT	COMMENTS
Methane (CH_4) (Marsh gas)	Gas	999° F. 537°C.	5.3	14.0		0.6	−259° F. −162°C.	Colorless, odorless, asphyxiating gas. Chief component of natural gas. Used widely as a fuel and a source of carbon black and simple organic chemicals. Red label.
Ethane (C_2H_6)	Gas	959° F. 515°C.	3.0	12.5		1.0	−128° F. −89°C.	Colorless, odorless, asphyxiating gas. Component of natural gas. Used as a refrigerant and fuel, and in the manufacture of organic chemical compounds.
Propane (C_3H_8)	Gas	871° F. 466°C.	2.2	9.5		1.6	−44° F. −42°C.	Colorless, asphyxiating gas with little natural odor. One of the Liquefied Petroleum (LP) gases. Also used as refrigerant, solvent, organic synthesis.
Butane (C_4H_{10})	Gas	761° F. 405°C.	1.9	8.5		2.0	31° F. −0.6°C.	Colorless, asphyxiating gas with little natural odor. One of the Liquefied Petroleum (LP) gases. Also used in chemical manufacture and as refrigerant.
Pentane (C_5H_{12})	−40° F.* −40°C.*	588° F. 309°C.	1.5	7.8	0.6	2.5	97° F. 37°C.	Colorless liquid with a pleasant odor. Used as a solvent, as an anesthetic, in artificial ice manufacture. Petroleum ether is partly composed of pentane.

TABLE 3–2. PARAFFIN SERIES (CONT.)

	FLASH POINT	IGNITION TEMP.	LOWER FLAMMABLE LIMIT (% BY VOL. IN AIR)	UPPER FLAMMABLE LIMIT (% BY VOL. IN AIR)	SPECIFIC GRAVITY (WATER=1.0)	VAPOR DENSITY (AIR=1.0)	BOILING POINT	COMMENTS
Hexane (C_6H_{14})	−7° F. −28°C.	453° F. 234°C.	1.2	7.5	0.7	3.0	156° F. 69°C.	Colorless liquid with a faint odor. Used as a solvent and as a liquid in low temperature thermometers. Petroleum ether is partly composed of hexane.
Heptane (C_7H_{16})	25° F. −4°C.	433° F. 223°C.	1.2	6.7	0.7	3.5	209° F. 98°C.	Colorless liquid used as a solvent, as an anesthetic, and in the preparation of chemicals and reagents. One of the principal components of gasoline.
Octane (C_8H_{18})	56° F. 13°C.	428° F. 220°C.	1.0	3.2	0.7	3.9	258° F. 126°C.	Colorless liquid used as a solvent and in the manufacture of chemicals. One of the standards for gasoline rating, an isomer of octane is a principal component of gasoline.
Nonane (C_9H_{20})	88° F. 31°C.	403° F. 206°C.	0.8	2.9	0.7	4.4	303° F. 151°C.	Colorless liquid used in the manufacture of organic chemicals. One of the components in gasoline, kerosene, and naphtha.
Decane ($C_{10}H_{22}$)	115° F. 46°C.	406° F. 208°C.	0.8	5.4	0.7	4.9	345° F. 174°C.	Colorless liquid used as a solvent and in the manufacture of organic chemicals. One of the principal components of naphtha and kerosene.

*Minus 40°F. and minus 40°C. are the only points on the two temperature scales where the readings are identical.

About an hour after the original fire began, the top of the spheroid ruptured, releasing a huge ball of burning vapors with only the slightest warning to the fire fighters. All 19 were killed immediately, and their bodies were found in an area between 300 and 400 feet from the tank. The ball of vapors also ignited three crude oil tanks, 450 to 550 feet away. Spectators, standing on a highway 1200 feet away, were burned. A railroad trestle 1250 feet away was destroyed. Paint blistered on houses 3000 feet away, and leaves burned on trees a mile away. (For a complete report see NFPA *Fireman's Magazine,* September, 1956.)

This label, with black lettering on a red background, is required when any flammable liquid is shipped. An additional placard, shown on page 50, is required when 1000 pounds (454 kilograms) is to be shipped.

FIGURE 3–2. DOT Flammable Liquid Labels

Gasoline Gasoline is a typical liquid hydrocarbon. It burns vigorously with a bright orange flame and quantities of black smoke. It is lighter than water and is not miscible in water. Although not considered highly toxic,

The placard must be red with a white symbol, inscription, and ½-inch (12.7-mm) border. The word *GASOLINE* may be used in place of the word *FLAMMABLE* on the placard that is displayed on a cargo tank or portable tank being used to transport gasoline by highway. The word *GASOLINE* must be in letters of the same size and color as those in the word *FLAMMABLE*.

FIGURE 3–2. DOT Flammable Liquid Labels (cont.)

gasoline can quickly build up a lethal concentration of carbon monoxide when burning in an enclosed space in or out of the engine of an automobile. It is also a skin irritant. Many mechanics have "strawberries" from carrying a gasoline- or kerosene-soaked rag in a back pocket.

Structurally, gasoline is a blend of ISOMERS. Sometimes, the same number of atoms can hook together in a slightly different manner. This revised lineup is called an isomer. Both octane and iso-octane have the same molecular formula: C_8H_{18}. Structurally, however, they are not exactly alike.

Modern gasolines are blended from a wide variety of hydrocarbons (and their isomers) ranging from pentane to nonane. Occasionally, benzene,

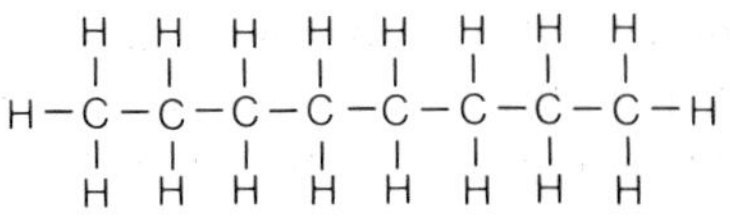

```
    H   H   H   H   H   H   H   H                H   H   H   H   H   H   H
    |   |   |   |   |   |   |   |                |   |   |   |   |   |   |
H - C - C - C - C - C - C - C - C - H        H - C - C - C - C - C - C - C - H
    |   |   |   |   |   |   |   |                |   |   |   |   |   |   |
    H   H   H   H   H   H   H   H                H   H   H   C   H   H   H
                                                            /|\
                                                            HHH
```

Octane
C_8H_{18}

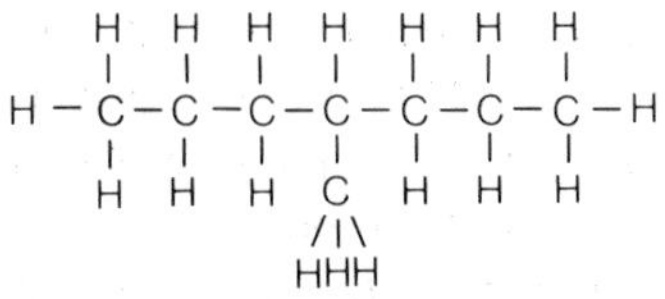

Iso-Octane*
C_6H_{18}

*Molecules such as iso-octane are called branched hydrocarbons

FIGURE 3–3. Isomers Used in Gasoline

toluene, and certain alcohols are included in the blends. For fire fighters, the importance of the difference between gasolines is their somewhat different properties. One-hundred-octane gasoline has a slightly higher flash point and ignition temperature than lower octane gasoline. This can have some bearing on the types of sources which will ignite a particular gasoline vapor. Nevertheless, the degree of hazard for both types is very high. (See Table 3–1).

Misuse of gasoline constitutes one of the major fire hazards in our country today. Of necessity, gasoline is universally available. Besides being a fuel, gasoline is used as a solvent in certain industrial processes. As a consequence of its familiarity it is treated with a measure of contempt, but it is an extremely dangerous, flammable liquid. It is ready to burn or explode at any time. The misuse of gasoline for washing automotive parts or for dry cleaning creates an explosive air mixture that only requires a source of ignition. All too often, this ignition source is around the corner in the form of a water heater or stove pilot flame or the spark from an electric switch or motor. Tragically, thousands of people are killed or injured every year because of their ignorance of the lethal qualities of this familiar liquid.

Educating the public about misuses of gasoline is a major concern of the fire service. Misrepresentations, such as the final scene in a recent movie, where the heroine and the villain lit matches in a small darkened room and playfully tossed them at each other's gasoline-soaked clothing, do not help. The audience came away with the dangerous misconception that people in this situation would be safe as long as they kept dodging the matches. Hollywood was not concerned about the explosive limits of gasoline vapor.

Kerosene (Kerosine, Range Oil, Fuel Oil #1, Coal Oil) Kerosene is a mixture of hydrocarbons having ten to sixteen carbon atoms in their molecules (decane to hexadecane). It is a pale yellow to white liquid, with a strong, familiar odor. Skin-irritating qualities are about the same as gasoline and, as with all hydrocarbons, carbon monoxide can be a product of incomplete combustion. Kerosene is formulated to have a flash point between 100°F. (38°C.) and 140°F. (60°C.). With a flash point in this range, it will not produce flammable vapors at ordinary temperatures. Used as a cleaner or degreaser, it is much safer than gasoline. However, if kerosene is heated to its flash point, the vapors are readily ignited and its heat of combustion is high. Kerosene burns hot, and its fire potential when exposed to heat should not be underestimated.

Kerosene was once the most economically important fraction of petroleum. Homes were lit by kerosene lamps and heated by coal oil stoves. Although electrification and natural gas have greatly reduced the use of kerosene for these purposes, portable kerosene heaters are still a major cause of loss of life by fire and asphyxiation.

> Portable kerosene heaters are not required to be connected to a chimney or flue. Since air is required for proper combustion, the appliance should not be used for long periods of time in a room of tight construction. Under such circumstances, the oxygen in the air may become exhausted. Improper combustion of the fuel oil results in the liberation of carbon monoxide. Therefore, the appliance should not be used in sleeping quarters and, when in use in other rooms, a window or door to the outside should remain partly open.
>
> There is also the common fault of leaving these appliances unattended after lighting. As the burner is heated, the vapor release increases. This results in a flame that increases in height and vapor release. It is not good practice to light the appliance, turn the control to the operating position and then leave it without further attention.*

Kerosene is also widely used as a cleaner, degreaser, and solvent. In recent years, it has staged a comeback as a fuel. Military and commercial jet aircraft generally burn kerosene (Commercial type "A") or a gasoline-kerosene mixture (Commercial type "B"). JP-4, containing 65 percent gasoline, has a somewhat higher flash point, −10°F. to 30°F. (−23°C. to −1°C.), and a lower ignition temperature, 468°F. (242°C.) than gasoline. Some consequences of these differences are described in the following quotation from AIA Special Interest Bulletin 182: "Jet Aircraft Fuels."

> Since the jet fuels have somewhat lower volatility than gasoline, if they are spilled and ignited, the resulting fire may not spread across the surface quite as rapidly as with gasoline. This is particularly true with kerosene-type fuels. In a major airplane crash, however, the impact is sufficient to atomize the spilled fuel. This fuel mist will

*From AIA Special Interest Bulletin 181, "Portable Kerosene Heaters," revised June 1956.

ignite readily, propagating flame rapidly in spite of any theoretical advantage of lower volatility in kerosene-type fuels.

Since water will cool kerosene below its flash point, the use of water fog is recommended. Ordinary foam, carbon dioxide, and dry chemical will also extinguish kerosene fires. Table 3–1 includes a summary of the characteristics of kerosene.

Naphtha To a perplexed fire fighter lost in a jungle of chemical terminology, there seems to be some truth in the story about a convention of chemists sitting around inventing names for chemicals: "Gentlemen," said one scientist, "here we have fifty completely different, new hydrocarbons. Let's call 'em all naphtha!"

Naphtha seems to be everybody's favorite name for a liquid. There are naphthas with high flash points and low flash points. There are heavy naphthas and light naphthas. There are naphthas made from petroleum and completely different naphthas made from coal tar. "Stoddard solvent" and "safety solvent" are both called high flash naphtha or petroleum spirits, or thinner, or mineral turpentine. Lighter fluid is sometimes called petroleum naphtha when it is not being called petroleum ether (which also has a couple of additional names). Varnish makers and painters use a lot of naphtha. There are at least three kinds of "painter's naphtha," each with a different set of characteristics and several names. Sorting out naphthas according to hazard, therefore, is difficult. NFPA Pamphlet 325 recognizes this in a little note: "Flash point and ignition temperature (of this naphtha) will vary, according to the manufacturer."

About this time, the student thinks longingly of a simple, straightforward compound like ethyl alcohol. But, if the fire fighter wants to know something more about naphtha, Table 3–1 lists some common examples.

Petroleum Ether The name of this liquid also causes trouble. Chemically, petroleum ether is not really an ether, so many chemists object to calling it one. Petroleum ether is often called benzine, but there are objections to this name too. It can be confused with benzene, a different hydrocarbon, so the use of the word *benzine* is being actively discouraged, although there is an official grade of this liquid called petroleum benzin. LIGROIN is the correct name for petroleum ether. Hardly anyone uses it, except in the laboratory. As we might guess, ligroin is sometimes called petroleum naphtha, to add to the confusion.

Whatever you call it, petroleum ether is a clear, colorless, highly flammable liquid with a strong odor. Depending on its manufacturer, its flash point ranges upward from −40°F. (−40°C.) to around 0°F.(−17°C.). Like all hydrocarbons, it is insoluble in water. Petroleum ether is the lightest liquid fraction of petroleum. Made from pentanes and hexanes, it is used as a solvent in the laboratory; as a solvent in drugs, oils, fats, and waxes; as a fuel; in paints and varnishes; as an insecticide; in photography; and as a detergent. Use as a spot remover and as a dry cleaner has been discouraged because of its flammability. Ligroin is often shipped and stored in one-gallon (four-liter) glass containers.

Inhaling petroleum ether vapors may cause headache, nausea, and intoxication, with consequent loss of judgment. Fire fighters likely to be exposed should wear self-contained masks and protective clothing. Extinguishment of petroleum ether fires with water is a questionable procedure. Use dry chemical or ordinary foam.

Fuel Oils There are many different kinds of fuel oils in use today. Fuel Oil No. 1 is a synonym for kerosene or range oil; No. 1D is a light diesel fuel. Fuel Oil No. 2 is used as a domestic fuel oil; No. 2D is a medium diesel fuel; No. 3 is no longer a current standard. Numbers 4, 5, and 6 are light, medium, and heavy industrial fuels, respectively. These oils become increasingly viscous from No. 1 through No. 6 and different fuel oil burners are designed for oils of particular viscosities.

VISCOSITY is the resistance offered by a liquid to flow. Syrup, for example, is more viscous than water. Acetone is less viscous than water, more mobile, although only a scientist would notice the difference. Several factors affect the viscosity of fluids. The colder a viscous liquid, the slower it pours: "as slow as molasses in January." And, as the length of a hydrocarbon chain increases, the viscosity of the resulting liquid also increases.

Viscosity assumes fire-fighting significance in the case of fuel oils. A specialized fire-fighting method can be used against heavy grade fuel oils, asphalt, and lubricating oils. If oil and water are shaken together, tiny droplets will form on the surface of the liquid. Although the water and the oil have not mixed, water is trapped inside these bubbles. This is called an EMULSION. Directing a coarse water spray onto the surface of a viscous liquid whose vapors are burning forms an emulsion that can smother the fire. This emulsion, while temporary in light fuel oils, is very persistent when formed with heavy grade oils. It will even guard against the possibility of a flashback.

The emulsification technique must be used with care because there is danger in using water in fires involving high-flash-point viscous liquids like these. Most have a flash point above 212°F. (100°C.). The heated liquid, of course, is above that temperature. If water is trapped underneath the surface, it will turn to steam, increase its volume, and cause boilovers, steam explosions, or violent spattering. Naturally, water can lower the temperature of these liquids below their flash points. But a great deal of heat is present. These liquids were hot to begin with, and their heats of combustion are high.

It may require great quantities of water to extinguish them, presenting the dangers of trapped water. Some fire scientists advocate the circulation of cooler oil from the bottom of the tank, lowering the temperature of the surface liquid below its flash point, as a possible method of extinguishment. Foam has also been used.

Most of these comments on fuel oils also apply to lubricating oils and asphalt. (See Table 3–1 for a summary of the properties of these liquids.)

Ring Hydrocarbons

Besides chains and branched chains, other types of hydrocarbon series are possible. The ever-versatile carbon atom is also capable of forming rings. Three examples of ring hydrocarbons are *benzene, toluene,* and *xylene.* Being hydrocarbons, they are flammable, but they have an added hazard: a rather high level of toxicity.

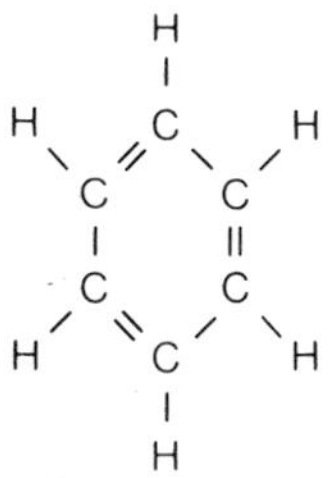

```
            H
            |
    H       C       H
     \   //   \   /
       C         C
       |         ||
       C         C
     /   \\   /   \
    H       C       H
            |
            H
```

FIGURE 3–4. Benzene (C_6H_6)

Benzene (Benzol) Benzene, the most toxic of the ring hydrocarbons, is very widely used in industry. It is used in the manufacture of styrene, phenol, and nylon. Aniline and other dyes are made from benzene. It is used in the production of such insecticides as DDT and benzene hexachloride (Lindane); to make many organic compounds; in the manufacture of medicinal chemicals, artificial leathers, and oilcloth; and as a solvent in varnishes, lacquers, resins, waxes, oils, and such commodities as airplane dope. Large amounts of benzene are used in motor fuels, particularly aviation and 100+ octane gasolines.

Benzene can damage the bone marrow, producer of the body's blood cells, causing anemia. It is also suspected of producing leukemia (a blood cancer) and of affecting the liver and kidneys. Symptoms of poisoning may not appear until the effect on the body has approached fatal proportions. Benzene has a cumulative action, and prolonged inhalation, such as by workmen constantly exposed to its fumes, is considered the most dangerous type of exposure. *Brief high exposures, the major hazard to fire fighters, can cause anesthesia, leading to death by respiratory failure.*

Benzene vapors are insidious. They have a pleasant smell and give no adequate warning of their toxic effects. Because of their pleasing smell, members of the ring series are often called the *aromatic hydrocarbons*. There is no appreciable irritation or discomfort when we inhale benzene, even at dangerously toxic levels. The liquid can be absorbed through the skin in harmful amounts, causing irritation and even blisters. Symptoms of benzene poisoning include dizziness, a tightening of the leg muscles, forehead pressure, intoxication, and eventual coma. Fire fighters should protect themselves from aromatic hydrocarbon vapors with self-contained breathing apparatus.

Benzene has an extremely high concentration of carbon in its molecule, as compared to other hydrocarbons. Benzene (and other members of the ring series) burn with an extremely smoky flame, composed of particles of unburned carbon escaping from the combustion zone, and produce large quantities of carbon monoxide. Although fires involving this clear, colorless liquid can be fought with water spray, this may not be the best procedure. Dry chemical, CO_2, or ordinary foam may give better results. Class 1, Group D, electrical equipment should be used wherever benzene vapors may be present. (See Chapter 5, page 115 for classifications.)

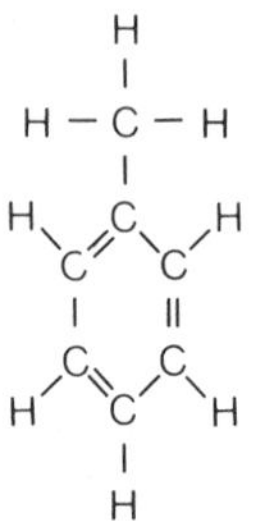

```
        H
        |
    H - C - H
        |
  H     C     H
    \ //  \  /
     C     C
     |     ||
     C     C
    / \\  / \
  H     C     H
        |
        H
```

FIGURE 3–5. Toluene (C_7H_8)

Toluene (Toluol) Toluene is simple benzene with an added carbon side chain, as shown. This side chain is known as a METHYL GROUP, and toluene is therefore sometimes called *methylbenzene*. Along with benzene, toluene is generally produced from coal. Toluene is a flammable, colorless liquid with an easily detected benzene odor. It is used in the manufacture of organic compounds; in lacquers, aviation gasolines, medicines, dyes, perfumes, and saccharin. As a solvent, it will dissolve gums, resins, some oils, cellulose acetate, and cellulose ether. Trinitrotoluene (TNT) is a nitrated form of toluene. Toluene is shipped in containers ranging from glass bottles to large tanks.

Although it is an eye and respiratory irritant, and although extreme inhalation of vapors may cause paralysis of the respiratory center, acute poisonings from toluene are rare. It is less toxic than benzene but can cause fatigue, weakness, and loss of judgment.

Toluene fires can be fought with water spray although, once again, extinguishment is questionable. Better results may be obtained from the use of dry chemical, CO_2, or ordinary foam. Class 1, Group D, electrical equipment is necessary around toluene. Table 3–1 lists toluene's fire characteristics.

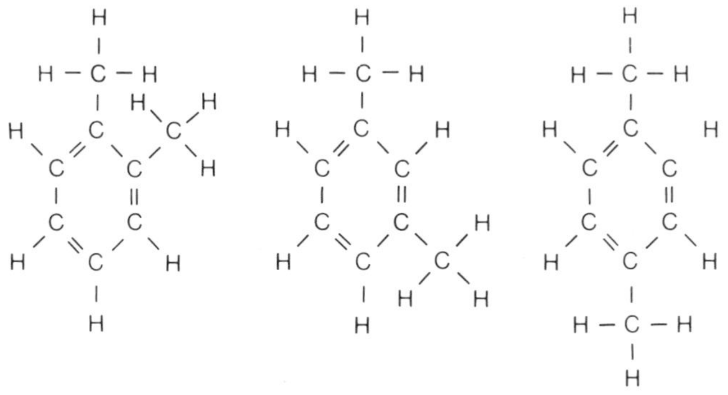

Ortho-xylene C_8H_{10} **Meta-xylene** C_8H_{10} **Para-xylene** C_8H_{10}

FIGURE 3–6. Isomers of Xylene

Xylene (Xylol) Toluene is benzene with an added carbon side chain. *Xylene* has two side chains which can be attached at different places* (as shown in Figure 3–6).

All of the xylenes are clear, colorless, flammable liquids. Their commonest use is in aviation gasoline, but xylenes will dissolve lacquers, enamels, and rubber cements. Other uses include the manufacture of dyes, vitamins, medicines, and polyester plastics. Xylenes are less toxic than benzene and give warning of their presence by irritating the eyes, nose, and throat. Vapors in high concentrations are anesthetic, and a few cases of fatal poisoning have occurred. Like the other ring hydrocarbons, xylene burns with an extremely smoky flame, producing carbon monoxide in quantity. Fire-fighting methods parallel those for benzene and toluene. See Table 3–1 for a summary of xylene fire characteristics.

---

*Xylene has 3 isomers: ortho-xylene (*o*-xylene), para-xylene (*p*-xylene), and meta-xylene (*m*-xylene), known respectively as 1,2 xylene, 1,3 xylene, and 1,4 xylene. Count the carbon atoms in the benzene ring clockwise to see why.

Turpentine Turpentine is a major liquid hydrocarbon not derived from either coal or oil. It is not a ring hydrocarbon. Crude turpentine is a sticky yellowish gum obtained from pine trees. From this balsam oil of turpentine is distilled. Oil or spirits of turpentine is a colorless liquid with a characteristic odor. It is used as a solvent for oils, resins, varnishes, paints, lacquers, rubbers, polishes, and waxes.

Excessive exposure to turpentine fumes can cause nausea and headache. Kidney injury is possible. Its fumes are irritating to the nose and throat. Fire fighters should be protected by self-contained masks, since burning turpentine produces carbon monoxide.

OIL OF TURPENTINE, unlike most hydrocarbons, heats spontaneously. It reacts vigorously with chlorine and iodine. Its fire characteristics are similar to those of kerosene.

FLASH POINT	IGN'T TEMP	FLAMMABLE LOWER	LIMITS UPPER	SPECIFIC GRAVITY	VAPOR DENSITY	BOILING POINT	WATER SOLUBLE
95° F. 35°C.	488° F. 253C.	0.8	. . .	1.0	4.8	300° F. 149°C.	No

Alcohol

When discussing combustion, we mentioned that molecule fragments in a combustion zone were called RADICALS. A particular group of atoms inside a molecule that tends to operate as a unit is also called a radical. Notice the O–H combination on the right side of the structural formula of METHYL ALCOHOL (Figure 3–7). This O–H combination is known as the HYDROXIDE (or hydroxyl) RADICAL. When a hydrocarbon adds a hydroxyl radical, it generally changes into an alcohol. Indeed, the combination C–O–H means "alcohol" to a chemist.

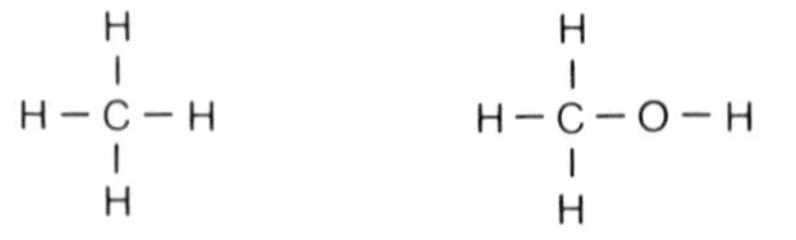

```
    H                 H
    |                 |
H - C - H         H - C - O - H
    |                 |
    H                 H
```

Methane
(CH_4)

Methyl alcohol
(CH_3OH)

FIGURE 3–7. Methyl Alcohol

When an O–H group is added to an inorganic element or group of elements that does not include a carbon atom, the resulting substance is called a *base*. NaOH, called sodium hydroxide or lye, is a very caustic basic (alkaline) material. NH_4OH is ammonium hydroxide, a weak solution of NH_4OH is called ammonia water. These compounds *are not* alcohols.

Stripped of the hydroxyl radical, the left side of the methyl alcohol molecule looks like this:

```
    H
    |
H – C –
    |
    H
```

FIGURE 3–8. The Methyl Radical

This group of four atoms is known as the METHYL RADICAL. We saw that when benzene added this group of atoms to its molecule, it changed into toluene. Whenever this combination appears as part of a larger molecule, it is generally reflected in the title of the compound: methyl alcohol, methyl acetate, methyl ether, and so forth.

When other members of the paraffin series are stripped of one of their hydrogen atoms, they also form radicals. When they add a hydroxyl radical, they change into alcohols (see Table 3–3).

TABLE 3–3. NAMES OF HYDROCARBONS, RADICALS, AND ALCOHOLS

HYDROCARBON	RADICAL	ALCOHOL	OFFICIAL NAME
Methane	Methyl	Methyl alcohol	Methanol
Ethane	Ethyl	Ethyl alcohol	Ethanol
Propane	Propyl	Propyl alcohol	Propanol
Isopropane	Isopropyl	Isopropyl alcohol	Isopropanol
Butane	Butyl	Butyl alcohol	Butanol
Isobutane	Isobutyl	Isobutyl alcohol	Isobutanol
Pentane	Amyl*	Amyl alcohol	Pentanol
Hexane	Hexyl	Hexyl alcohol	Hexanol
Heptane	Heptyl	Heptyl alcohol	Heptanol
Octane	Octyl	Octyl alcohol	Octanol
Nonane	Nonyl	Nonyl alcohol	Nonanol
Decane	Decyl	Decyl alcohol	Decanol

*The five-carbon radical is more often called amyl than pentyl even though the official name for amyl alcohol is pentanol. You would also expect that benzene would form the benzyl radical. It doesn't. Toluene does. Benzene forms the phenyl radical.

Correct chemical terminology reserves the use of the suffix *-ol* for alcohols. This is why the names *benzol, toluol*, and *xylol* are technically incorrect, although they are frequently used. Benzene, toluene, and xylene are hydrocarbons, not alcohols.

As a group, alcohols generally have lower heats of combustion than equivalent hydrocarbons. Alcohols are partially oxidized, partially burned already. But, as their carbon chain grows longer, so does their heat of combustion. Amyl alcohol burns hotter than methyl alcohol. Both burn hotter than ordinary combustibles. A lengthening carbon chain changes other properties of alcohols. Methyl, ethyl, and the propyl alcohols are completely soluble in water. (The hydroxyl radical, O–H, is very similar to the structure of water, H–O–H.) Beginning with *butyl alcohol,* however, the water solubility of alcohols grows less and less as the hydrocarbon radical assumes greater weight. *Heavier alcohols may be totally insoluble in water.* Eventually, as the hydrocarbon chain grows longer, alcohols become heavier and heavier liquids until, at last, solid alcohols are formed. CETYL ALCOHOL, with a sixteen-carbon chain, is a white crystalline powder first discovered in a whale.

There are three major types of alcohols. The addition of a single hydroxyl group forms the alcohols listed in Table 3–3. Two hydroxyl groups attached to a chain form glycols. Three hydroxyl groups form glycerol (see Figure 3–9).

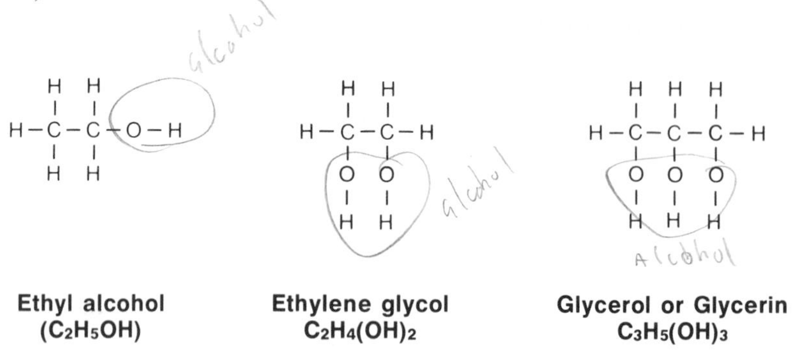

FIGURE 3–9. Alcohol and Related Compounds

Methyl Alcohol (Methanol, Wood Alcohol, Wood Spirits)

The production of methyl alcohol in the United States is over 150,000,000 gallons annually. In terms of widespread use, it is one of the "big three" alcohols, along with ethyl alcohol and isopropyl alcohol. Were it not for its toxicity, methyl alcohol would be used even more.

Once prepared by heating wood in the absence of air (hence, its name "wood alcohol"), 90 percent of all methyl alcohol is now made by the partial oxidation of methane or by combining carbon monoxide and two hydrogen molecules, under heat and pressure, with the aid of catalysts, zinc oxide and chromic acid. (A CATALYST is a material which speeds or assists a chemical reaction but takes no part in the reaction. Catalysts are frequently employed in industrial processes.)

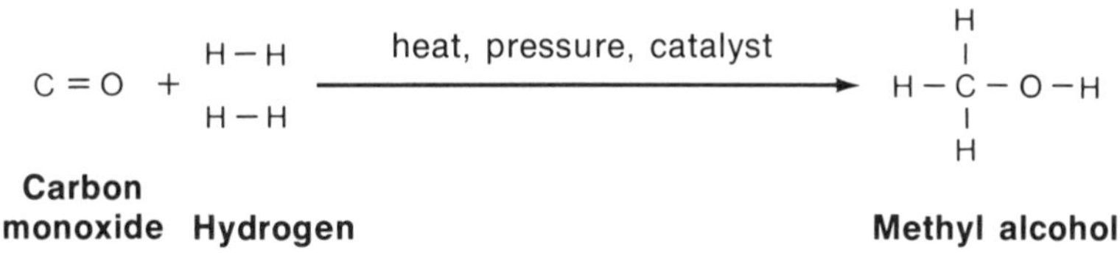

$$\underset{\text{Carbon monoxide}}{C{=}O} + \underset{\text{Hydrogen}}{\begin{matrix} H-H \\ H-H \end{matrix}} \xrightarrow{\text{heat, pressure, catalyst}} \underset{\text{Methyl alcohol}}{\begin{matrix} & H & & \\ & | & & \\ H- & C & -O- & H \\ & | & & \\ & H & & \end{matrix}}$$

FIGURE 3–10. Preparation of Methyl Alcohol

Methyl alcohol is a clear, colorless, volatile, flammable liquid. When pure, it has a slight alcohol odor. Impure, the odor may become stronger. It burns with the typical blue flame of alcohols. Table 3–4 (pages 62–63) lists the fire characteristics of methyl alcohol. Since it is water soluble, dilution techniques can be used for extinguishment. In addition, water fog, in conjunction with dry chemical, ''alcohol'' foam, and carbon dioxide have all been used successfully.

Deaths have occurred from drinking small amounts of methyl alcohol. Inside the body, it is oxidized into formaldehyde and formic acid, both of which are poisonous, and it is eliminated slowly. Inhalation of methyl alcohol vapors is also hazardous. While single exposures to fumes in low concentrations may cause no harmful effects, heavier concentrations can be dangerous. Depending on the amount of exposure, poisoning may cause a variety of symptoms: headache, fatigue, nausea, acidosis, dizziness, intoxication, unconsciousness, a sighing respiration, respiratory failure, and circulatory collapse.

The greatest danger from methyl alcohol is its specific toxic effect upon the eyes. It can atrophy the optic nerve and retina. Visual symptoms, such as blurring of vision, clear up, only to recur and progress to blindness. Exposure to methyl alcohol liquid may cause the skin to become dry and cracked, and poisoning can occur through breaks in the skin.

Flammable liquids are mainly used as fuels and solvents. Methyl alcohol is a better solvent than ethyl alcohol. It is abundant and cheap, and has a low boiling point so it can be easily vaporized after it is no longer needed. As a consequence, in spite of its toxicity (it cannot be made nontoxic), methyl alcohol is widely used in industry. It is a solvent for surface coatings of various kinds, for paint removers, inks, and adhesives. It is used in the production of formaldehyde and the synthesis of various chemicals. It is sometimes used as an antifreeze in gasolines and diesel oils where it combines with the water in these fuels to prevent the formation of ice. Other

TABLE 3–4. PROPERTIES OF VARIOUS ALCOHOLS, GLYCOLS, AND GLYCEROL

	FLASH POINT	IGNITION TEMP.	FLAMMABLE LIMITS (PERCENT BY VOLUME IN AIR)		SPECIFIC GRAVITY (WATER=1.0)	VAPOR DENSITY (AIR=1.0)	BOILING POINT	WATER SOLUBLE
			LOWER	UPPER				
Allyl alcohol* (Propenyl alcohol)	70° F. 21°C.	713° F. 378°C.	2.5	18.0	0.9	2.0	206° F. 97°C.	Yes
U.L. Class 40: Amyl alcohol (normal) (Pentanol)**	91° F. 33°C.	572° F. 300°C.	1.2	10.0	0.8	3.0	280° F. 138°C.	Slightly
U.L. Class 40: Butyl alcohol (Butanol) (normal)	84° F. 29°C.	650° F. 343°C.	1.4	11.2	0.8	2.6	243° F. 117°C.	Yes
(secondary)	75° F. 24°C.	763° F. 406°C.	1.7	9.8	0.8	2.6	201° F. 94°C.	Yes
(tertiary)	52° F. 11°C.	892° F. 478°C.	2.4	8.0	0.8	2.6	181° F. 83°C.	Yes
U.L. Class 70: Denatured alcohol***	60° F. 16°C.	750° F. 399°C.			0.8	1.6	175° F. 79°C.	Yes
U.L. Class 70: Ethyl alcohol**** (Ethanol, Grain alcohol)	55° F. 13°C.	793° F. 423°C.	4.3	19.0	0.8	1.6	173° F. 78°C.	Yes
Ethylene glycol (Glycol)	232° F. 111°C.	775° F. 413°C.	3.2		1.1		387° F. 197°C.	Yes
U.L. Class 10-20: Glycerol (Glycerin)	320° F. 160°C.	739° F. 393°C.			1.3		554° F. 290°C.	Yes

TABLE 3–4. PROPERTIES OF VARIOUS ALCOHOLS, GLYCOLS, AND GLYCEROL (CONT.)

	FLASH POINT	IGNITION TEMP.	FLAMMABLE LIMITS (PERCENT BY VOLUME IN AIR)		SPECIFIC GRAVITY (WATER=1.0)	VAPOR DENSITY (AIR=1.0)	BOILING POINT	WATER SOLUBLE
			LOWER	UPPER				
U.L. Class 35-40: Isoamyl alcohol***** (Fuel oil)	109° F. 43°C.	657° F. 347°C.	1.2	9.0 (at 212° or 100°C.)	0.8	3.0	270° F. 132°C.	Slightly
U.L. Class 40-45 Isobutyl alcohol (Isopropyl carbinol)	82° F. 28°C.	800° F. 427°C.	1.7 (at 212°F.	10.9 or 100°C.)	0.8	2.6	225° F. 107°C.	Yes
U.L. Class 70: Isopropyl alcohol (Isopropanol)	53° F. 12°C.	750° F. 399°C.	2.1	13.5	0.8	2.1	181° F. 83°C.	Yes
U.L. Class 70: Methyl alcohol (Methanol, Wood alcohol)	52° F. 11°C.	867° F. 464°C.	7.3	36.0	0.8	1.1	147° F. 64°C.	Yes
U.L. Class 55-60: Propyl alcohol (Propanol)	77° F. 25°C.	700° F. 371°C.	2.1	13.5	0.8	2.1	207° F. 97°C.	Yes
Propylene Glycol (Methyl ethylene glycol)	210° F. 99°C.	790° F. 421°C.	2.6	12.5	1.0 (or over)		370° F. 188°C.	Yes

*Allyl alcohol has a pungent odor somewhat like mustard. It is poisonous and very dangerous to the eyes. Absorbed readily through the skin, highly toxic by inhalation or ingestion. Odor may not be sufficient to warn fire fighters to don protective equipment! Used in several industrial processes and as a military poison gas. Shipped under a poison "B" label.

**An isomer of amyl alcohol, with a lower boiling point of 245°F. (118°C.) has a classification of 40-45. This isomer has a flash point of 94°F. (34°C.).

***Different government formulations can lower the flash point to less than 20°F. (−7°C.).

****Water mixtures have higher flash points.

*****There are two other isomers of isoamyl alcohol, one of which has a flash point of 67°F. (19°C.).

uses include manufacture of medicinals, softener for pyroxylin plastics, and as a fuel in such appliances as picnic stoves. Methyl alcohol is shipped in a wide variety of containers.

Ethyl Alcohol (Ethanol, Grain Alcohol, Alcohol, Cologne Spirits)

Ethyl alcohol, mixed with water and flavorings, seems to be everybody's favorite flammable liquid. It is by far the most important industrial alcohol, so much so that the amount of ethyl alcohol sold is a direct reflection of the American economy. It is used to manufacture a variety of industrial compounds, such as aldehydes, acids, ethers, acetates, drugs and medicinals (tincture of iodine is iodine dissolved in ethyl alcohol), dyes, certain rubbers, and explosives. It is used in food flavorings and extracts, cosmetics, and lotions of various kinds. It is a solvent for resins and varnishes. Small wonder the general term *alcohol* has come to mean ethyl alcohol, although there are many different alcohols.

Sugars found in such substances as grain, fruits, and molasses, when added to water and yeast, will ferment into "grain alcohol," while giving off carbon dioxide.

$$\underset{\textbf{Sugar}}{C_{12}H_{22}O_{11}} + \underset{\textbf{Water}}{H_2O} \xrightarrow{\text{yeast}} 4\,\underset{\textbf{Ethyl alcohol}}{C_2H_5OH} + 4\,\underset{\textbf{Carbon dioxide}}{CO_2}$$

The flammable gas ETHYLENE is used in the two major industrial processes which form synthetic ethyl alcohol. The most important method adds a water molecule to ethylene through the use of heat, pressure, and a catalyst.

$$\underset{\textbf{Ethylene}}{H{-}\overset{\displaystyle H}{\overset{|}{C}}{=}\overset{\displaystyle H}{\overset{|}{C}}{-}H} + \underset{\textbf{Water}}{H{-}O{-}H} \xrightarrow{\text{heat, pressure, catalyst}} \underset{\textbf{Ethyl Alcohol}}{H{-}\underset{\displaystyle H}{\underset{|}{\overset{\displaystyle H}{\overset{|}{C}}}}{-}\underset{\displaystyle H}{\underset{|}{\overset{\displaystyle H}{\overset{|}{C}}}}{-}O{-}H}$$

Industrial processes produce pure ethyl alcohol, sometimes called "absolute alcohol" or ANHYDROUS (WATER-FREE) ALCOHOL, a colorless, volatile, flammable liquid with a familiar odor (although some absolute alcohols are odorless) and a burning, pungent taste. Absolute alcohol is so water soluble it absorbs water from the air. Since ethyl alcohol is totally water soluble, fire-

TABLE 3–5. DILUTION EFFECT ON FLASH POINT OF ETHYL ALCOHOL

PERCENTAGE OF ALCOHOL	PROOF	FLASH POINT	
		F.	C.
100	200	55°	13°
96	192	62°	17°
95	190	63°	17°
80	160	68°	20°
70	140	70°	21°
60	120	72°	22°
50	100	75°	24°
40	80	79°	26°
30	60	85°	29°
20	40	97°	36°
10	20	120°	49°
5	10	144°	62°

fighting methods like dilution can be very effective. As additional water is added to ethyl alcohol (about six parts to one), the flash point will rise above 100°F. (38°C.). This property is useful in diluting or extinguishing a spill. Adding sufficient water creates a mixture that is difficult to ignite (see Table 3–5). Dilution techniques are less effective on tanks, where they can cause an overflow and compound the fire problem.

Whiskey Fires The following summary of whiskey warehouse fire hazards was taken from the AIA Research Report #5, "Processes, Hazards and Protection Involved in the Manufacture of Spirituous Liquors."

Experience shows that if a fire gains considerable headway in an open rack whiskey warehouse, it cannot be extinguished. Because of the intense heat, large quantities of highly flammable vapors are immediately produced, causing fire to spread very rapidly, and making fire fighting exceedingly difficult.

An explosive mixture will not ordinarily exist more than a short distance from the source of vapor* because alcohol vapor diffuses readily. But, while alcohol is soluble in water, favoring the dilution method of attack, it is slightly lighter than water. Unless well mixed, a certain percentage will float on the water, especially when hot, causing streams of burning whiskey to endanger surrounding property.

Provided the fire is attacked promptly and control is achieved before the barrels start to contribute substantial quantities of flammable liquid to the fire, whiskey warehouse fires can be extinguished either by a properly designed automatic sprinkler system, or by water spray properly applied by

* Whiskey has a flash point of 82°F. (28°C.).

manual methods. Ordinary foam is not effective on alcohol fires; "alcohol" foam is effective on whiskey spills, if applied at the proper rate.

There is a wide difference in the hazards attending the storage of whiskey in wooden barrels and the storage of alcohol in metal drums. Under fire conditions, a barrel of whiskey in storage will not explode. Instead, after a period of time, it will begin to leak, eventually draining its entire contents into the fire and being itself consumed. On the other hand, if the usual uninsulated metal drums are exposed to burning liquid or fire in other nearby combustibles, they become especially dangerous and may burst with explosive violence, damaging other drums and adding more liquid to the original fire.

An unusual obstacle is that bonded whiskey warehouses, like distilleries, are kept securely locked from sundown to sunup and only one man, the government storekeeper, has the keys. Further, all window openings within ten feet of the ground are barred.

Forcing entry into buildings, either involved or exposed, has long been recognized as a right of a fire department or private fire brigade if circumstances require such action. This right of forcible entry is not restricted because a building is a bonded warehouse. Valuable time lost in the early stages of a fire, when every minute counts, may mean the difference between controlling and eventually extinguishing a small fire and the destruction of a large and valuable warehouse. Both the owner and the government have much at stake in such circumstances. When entrance by force is made, the proprietor must notify the storekeeper-gauger without delay, and maintain strict guard until his arrival.

The inhalation of ethyl alcohol vapors is anesthetic, but is not particularly toxic. Some eye irritation can occur at higher concentrations. The ingestion of ethyl alcohol can cause nausea, vomiting, flushing, drowsiness, toxication, or even coma or stupor. Drinking ethyl alcohol in sudden large amounts can cause respiratory collapse and death. However, unlike methyl alcohol, ethyl alcohol is oxidized fairly rapidly by the body into nontoxic products, carbon dioxide and water. There is no cumulative effect.

When ethyl alcohol is not used as a beverage or in cosmetics, various denaturants are added to it. Denatured alcohol is generally not taxable. Denaturants can include methyl alcohol, camphor, amyl alcohol, gasoline, isopropyl alcohol, turpentine, benzene, castor oil, acetone, nicotine, aniline dyes, ethers, sulfuric acid, kerosene, and other liquids not often recommended for drinking. These denaturants are generally chosen because they have boiling points similar to ethyl alcohol and cannot be easily separated by distilling. They can make pure ethyl alcohol more toxic and flammable.

Other Alcohols

To round out the description of the various alcohols, we should again mention ISOPROPYL ALCOHOL, rubbing alcohol, which also has important industrial uses. It is not considered toxic, although it can be locally irritating and, in high concentrations, can act as a narcotic.

ETHYLENE GLYCOL, widely used as a permanent antifreeze and motor coolant, is not considered highly toxic either, unless someone decides to drink it. A mixture of ethylene glycol and water will freeze at a lower temperature than either liquid separately.

GLYCERIN (glycerol) is often used in the making of candy, and the only health hazards it presents seem to be those associated with overeating.

Table 3–4 shows the fire characteristics of these three liquids.

Ethers

The smell of ether is familiar to anyone who has been in a hospital, and we are aware of the anesthetic properties of most ethers. Involved in an emergency, their vapors can produce unconsciousness (when such a state is least desirable) or even death from respiratory failure, if the exposure is severe.

DIOXANE, diethylene ether, has an irritating and poisonous vapor. Toxic amounts of liquid dioxane can be absorbed through the skin. Besides their use as anesthetics, ethers have a wide industrial value. Table 3–6 (page 68) lists some of them.

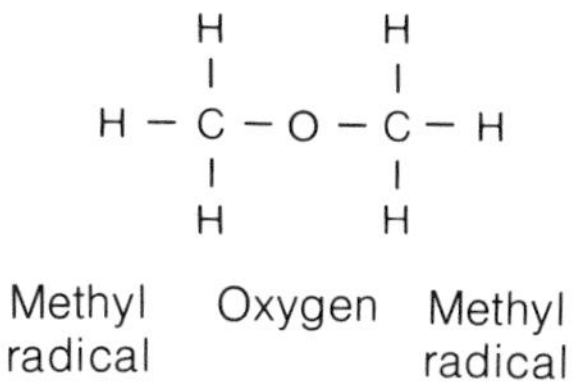

```
      H       H                 H   H       H   H
      |       |                 |   |       |   |
  H - C - O - C - H         H - C - C - O - C - C - H
      |       |                 |   |       |   |
      H       H                 H   H       H   H
```

Methyl radical — Oxygen — Methyl radical

Dimethyl ether
CH_3OCH_3 *

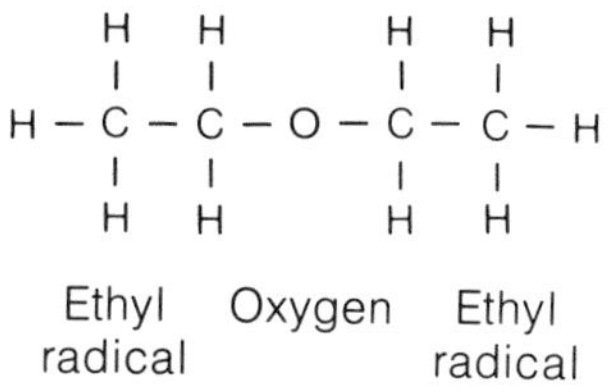

Ethyl radical — Oxygen — Ethyl radical

Diethyl ether
$C_2H_5OC_2H_5$

*Dimethyl ether and ethyl alcohol have exactly the same number of atoms of the same kind, two carbons, six hydrogens, and a single oxygen. (C_2H_6O). The difference in their properties is entirely due to the interior arrangement of the atoms.

FIGURE 3–11. Ether Molecules

As a class of flammable liquids, ethers present several specialized fire problems. Note the low boiling points of many. DIMETHYL ETHER (see Figure 3–11), which boils at −11°F. (−24°C.), is a flammable gas, and is only included here for comparison. METHYL ETHYL ETHER and ETHYL ETHER boil at such low temperatures that the heat of a man's hand can turn them into a gas. (This is *not* a recommended experiment.) As a consequence, these two ethers may be extremely difficult to extinguish. The amount of vapor they

TABLE 3–6. USES AND PROPERTIES OF ETHERS

	FLASH POINT	IGNITION TEMP.	LOWER FLAMMABLE LIMIT (% BY VOL. IN AIR)	UPPER FLAMMABLE LIMIT (% BY VOL. IN AIR)	SPECIFIC GRAVITY (WATER=1.0)	VAPOR DENSITY (AIR=1.0)	BOILING POINT	WATER SOLUBLE	COMMENTS
Butyl ether (Dibutyl ether)	77° F. 25°C.	382° F. 195°C.	1.5	7.6	0.8	4.5	286° F. 141°C.	No	Used as an industrial solvent. Has a mild odor. Avoid breathing vapor or skin contact with liquid.
Dimethyl ether (Methyl ether)	Gas	662° F. 350°C.	3.4	18.0		1.6	−11° F. −24°C.	Yes	Often shipped as liquid in pressurized cylinders. Uses: spray propellant, refrigerant, solvent. Requires DOT red label: fl. gas.
Dioxane—1, 4 (Diethylene oxide)	54° F. 12°C.	356° F. 180°C.	2.0	22.0	1.0	3.0	214° F. 101°C.	Yes	Irritating and poisonous vapor. Toxic amounts of liquid may be absorbed through the skin. Uses: industrial solvent, paint remover.
Ethyl ether (Diethyl ether; Diethyl oxide; Ethyl oxide; Ether)	−49° F. −45°C.	356° F. 180°C.	1.9	48.0	0.7	2.6	95° F. 35°C.	Slightly	Not corrosive or poisonous, but an anesthetic. Used as anesthetic, in manufacture of smokeless powder, as a solvent, in motor fuels, etc. Very low boiling point. Has a UL classification of 100.
Isopropyl ether (Diisopropyl ether)	−18° F. −28°C.	830° F. 443°C.	1.4	21.0	0.7	3.5	156° F. 69°C.	Slightly	Similar to ethyl ether in toxicity. Forms peroxides readily. Uses: solvent for oils, waxes, resins, dyes, to make rubber cements.
Methyl ethyl ether	−35° F. −37°C.	374° F. 190°C.	2.0	10.1	0.7	2.1	51° F. 11°C.	Yes	Note the boiling point and ignition temperature.
Vinyl ether (Divinyl ether)	−22° F. −30°C.	680° F. 360°C.	1.7	27.0	0.8	2.4	102° F. 39°C.	No	Wide anesthetic use. Can form peroxides. Low boiling point.

produce will be impressive. Their bubbling when heated, EBULLITION, may cause the breakdown of any foam blanket placed above them. Most ethers have a specific gravity less than water and will float upon it. Like alcohols, some are soluble in water, some are not; we have to identify the particular ether that confronts us.

The low ignition temperatures of many ethers means that the slightest ignition source can explode their vapors. This is why static electricity is such a hazard in hospital operating rooms and why potential ignition sources must be rigidly controlled. (We will cover this subject more fully when we investigate flammable anesthetics in Chapter 5.)

Ether Peroxides

Recently, the danger of ETHER PEROXIDE formation has received a great deal of publicity. There have been several incidents in which aging ether was discovered (particularly in colleges), taken to a supposedly safe place, and detonated. The resulting explosions were far greater than the initiating charge. On a few occasions, suspected bottles were taken to a vacant lot where stones were thrown at them. When struck, many of the bottles exploded. A doctor was killed when an ether bottle he was trying to open exploded and disemboweled him. Peroxide formation, the addition of oxygen atoms to the ether molecule, can be accelerated by several factors: heat, distillation, extended storage, and exposure to light or air. This is why most ethers are shipped in amber bottles or cans. Contact with certain metals, particularly iron and copper, appears to inhibit this formation and a piece of iron wire is often put inside ether cans. However, even in closed cans, formation of peroxides is possible. This is particularly true for isopropyl ether or absolute (pure) ether, although all ethers are suspect. Note that organic peroxides (ether peroxide is one of these) are generally classed as very unstable chemical compounds. (See Chapter 11 for further discussion.)

Storage and Shipping Conditions

The California State Fire Marshal's Office distributes the following excellent summary of safety measures for those who use ether.

1. Glass containers of all sizes should be avoided whenever possible.

2. All containers should be dated so that the age of the contents may be determined.

3. Isopropyl and absolute ethers should not be kept for more than six months; ethyl and other ethers for not more than one year.

4. Ether should be stored in as cool a location as feasible (but not stored in refrigerators unless explosion-proof).

5. Ether should always be tested for peroxide content before any distillation and, of course, should not be used if peroxides are found to be present.

6. Do not attempt to open any container of uncertain age or condition, or whose cap or stopper is tightly stuck.

7. Manufacturers should be contacted to learn any general recommendations regarding safe handling in storage and use, and any specific recommendations for the addition of inhibitors to prevent peroxide formation. Manufacturers can also recommend the best methods of chemical testing to detect peroxide content and the best methods of removing peroxides by chemical means.

Special precautions are necessary when ether is used as an anesthetic. As a matter of practice, ether is not used for this purpose if a can has been open longer than 24 hours. All of this brings to mind a disquieting picture of countless bottles and cans of ether, slowly turning dangerous on dusty rear shelves, in hospitals and schools around the country. Firemen called upon to dispose of a suspected container of ether should obtain technical help, although there are no reported cases of spontaneous explosion upon gentle handling.

Remember: Ether peroxides may not be visible as crystals. Treat any elderly ether as potentially explosive! After a fire, any bottles or cans exposed to heat should be discarded.

Finally, the name *ether* signifies to the chemist that the following lineup of atoms occurs in the molecule: carbon—oxygen—carbon. The various names given them—DIETHYL (two ethyls) ether or DIMETHYL (two methyls) ether—merely show how the entire molecule looks.

Amines

Several other important classes of flammable liquids remain. One of these, the amines, is a large group of colorless to yellow flammable liquids and gases, derived from ammonia, NH_3. When one or more of the hydrogen atoms in ammonia is replaced by a carbon-containing radical, an amine results (see Figure 3–12).

Most of the amines retain the aroma of ammonia, although a few smell more like rotten fish. Modern industry puts them to seemingly endless uses: in the manufacture of specialty soaps, dyestuffs, rubber chemicals, textile specialties, corrosion inhibitors, insecticides, drugs, wash and wear resins, fungicides, herbicides, petroleum additives, rubless floor polishes, shampoos, and so on.

As Table 3–7 shows, several of the amines present extraordinary hazards. ETHYL AMINE and ISOPROPYL AMINE have boiling points so low they turn to

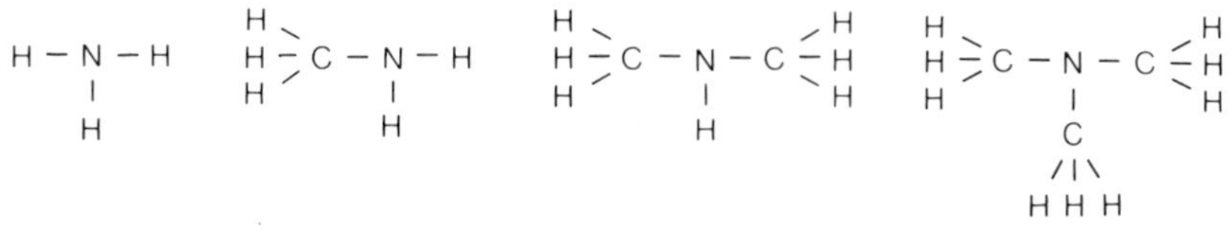

Ammonia NH_3 **Monomethylamine** CH_3NH_2 **Dimethylmine** $(CH_3)_2NH$ **Trimethylmine** $(CH_3)_3N$

FIGURE 3–12. Amine Formulas

gases at the slightest opportunity. Great care must be exercised when drums are opened. Refrigerated storage may be necessary. HYDROXYL AMINE explodes at 265°F. (129°C.). Remote fire-fighting methods and unmanned large appliances should be employed and the surrounding area evacuated.

Many of the amines are highly toxic. Some of the liquids will burn the skin, some will penetrate it. All of the vapors should be avoided by the use of self-contained gas masks.

Note: The Bureau of Mines Circular 8296, *Gas Masks for Respiratory Protection Against Amines,* discusses the difficulties in protecting against the amines' toxicity.

> Canister masks heat in proportion to the concentration of amine in the incoming air. Overheating indicates a dangerously high concentration and the wearer should return to fresh air. Even ammonia-type masks offer protection against only one or two of the amines.

Furthermore, anyone exposed to ETHYLENE DIAMINE can build up a sensitivity which renders him vulnerable to its effects. Do not give the material a second chance.

Precautionary labels on amines contain such language as: "Do not breathe vapor. Do not get in eyes, on skin, or on clothing. In case of contact with eyes or skin, or on clothing, immediately flush with plenty of water for at least fifteen minutes; for eyes get medical attention. Remove contaminated clothing and shoes at once, and wash thoroughly before re-use."

Aldehydes, Ketones, and Esters

The ALDEHYDES and the KETONES are chemically related. They both contain a carbon and an oxygen atom linked by a double bond. This is called a CARBONYL GROUP. Aldehydes and ketones are called carbonyl compounds.

Surrounding this double bond in the ketones are the various radicals which make each of them different. Acetone (dimethyl [two methyls] ketone) has the structure shown in Figure 3–13 (page 74).

TABLE 3–7. AMINES

	FLASH POINT	IGNITION TEMP.	FLAMMABLE LIMITS (% BY VOL. IN AIR)		SPECIFIC GRAVITY (WATER=1.0)	VAPOR DENSITY (AIR=1.0)	BOILING POINT	COMMENTS
			LOWER	UPPER				
Allyl amine	−20° F. −29°C.	705° F. 374°C.	2.2	22.0	0.8	2.0	128° F. 53°C.	Irritating and poisonous. Attacks rubber. Dangerous fire hazard.
Butyl amine	45° F. 7°C.	594° F. 312°C.	1.7	9.8	0.8	2.5	172° F. 78°C.	Liquid causes eye injury and skin irritation. Vapor harmful.
Cyclohexyl amine	90° F. 32°C.	560° F. 293°C.			0.9	3.4	274° F. 134°C.	Severe irritant, burns skin on contact. Vapors are nauseating.
Diethylene triamine	215° F. 102°C.	750° F. 399°C.			0.95		404° F. 207°C.	Danger! Liquid causes eye injury and skin burns. Vapor harmful.
Diisopropyl amine	30° F. −1°C.				0.7	3.5	183° F. 84°C.	Liquid may cause eye injury and skin irritation. Vapor harmful.
Ethyl amine	0° F. −18°C.	723° F. 384°C.	3.5	14.0	0.8	1.6	62° F. 17°C.	Extremely flammable, with a very low boiling point. Drums should not be opened in temperatures above 60°F. (16°C.) and opened slowly below that. Breathing of vapor very hazardous. Liquid causes eye injury and skin burns. This material has a severe fire hazard.
Ethylene diamine	110° F. 43°C.				9.0	2.1	241° F. 116°C.	Warning! Liquid causes eye injury and skin burns. Breathing of vapor may be harmful. Must carry a DOT poison label: Class B. Individuals exposed to this material may remain sensitive to it.
Hexyl amine	85° F. 29°C.				0.8	3.5	269° F. 132°C.	Only slightly water soluble. Liquid causes eye injury and skin burns. Can be absorbed through skin in harmful amounts.*

TABLE 3–7. AMINES (CONT.)

	FLASH POINT	IGNITION TEMP.	FLAMMABLE LIMITS (% BY VOL. IN AIR) LOWER	UPPER	SPECIFIC GRAVITY (WATER=1.0)	VAPOR DENSITY (AIR=1.0)	BOILING POINT	COMMENTS
Hydroxl amine (oxammonium)	Explodes at 265°F. (130°C.)				1.2		158° F. 70°C.	White crystals or a colorless liquid. Melts at 92°F. (33°C.) Use caution in approaching. Fight fires remotely. Moderate toxic hazards.
Methyl amine (mono)	Gas	806° F. 430°C.	4.9	20.7		1.1	21° F. −6°C.	Flammable gases. Liquid solution (25 percent to 40 percent methylamine in water) also flammable. Eye, skin, and respiratory irritants. Direct or sustained contact causes burns. Dry chemical or CO_2 extinguishers.
(di)	Gas	806° F. 430°C.	2.8	14.4		1.6	45° F. 7°C.	
(tri)	Gas	374° F. 190°C.	2.0	11.6		2.0	38° F. 3°C.	
Propyl amine (n)	−35° F. −37°C.	604° F. 318°C.	2.0	10.4	0.7	2.0	120° F. 49°C.	Extremely flammable. Breathing of vapor hazardous or fatal. Severe eye, skin, and respiratory irritant.
(iso)	−35° F. −37°C.	756° F. 402°C.			0.7	2.0	89° F. 32°C.	
Triamyl amine	215° F. 102°C.				0.8		453° F. 234°C.	Not water soluble. Eye, skin, and respiratory irritant.
Tributyl amine	187° F. 86°C.				0.8	6.4	417° F. 214°C.	Not water soluble. Eye, skin, and respiratory irritant.

*Skin absorption is also possible with triethylene tetramine and tetraethylene pentamine and other amines not listed in this table.

```
H \      / H
H - C - C - C - H
H /      ||      \ H
         O
```

FIGURE 3–13. Acetone (Dimethyl Ketone) (CH_3COCH_3)

In addition to acetone, methyl ethyl ketone (propanone), commonly called M.E.K., should be given special attention because of its wide use in industry as a degreasing and defatting agent (see Figure 3–14). This important property also makes it very irritating to the skin.

```
 H           H       H
   \         |     /
H - C - C - C - C - H
   /    ||   |     \
 H      O    H       H
```

FIGURE 3–14. Methyl Ethyl Ketone ($CH_3COC_2H_5$)

Other major ketones are widely used in industry as solvents for vinyl resins, esters, ethers, nitrocellulose lacquers, and pharmaceuticals, and to dewax lubricating oils. They assist in the manufacture of resins, dyes, plasticizers, and insecticides. Table 3–8 includes a summary of their fire characteristics. Their flammability is generally of more concern than their toxicity.

A spectacular fire involving storage tanks holding alcohols, ketones, and esters swept through a San Pedro Harbor loading dock area on August 8, 1972. The fire, described by one Los Angeles City Fire Department deputy chief as being "the worst chemical fire I've ever seen," caused an estimated $500,000 damage. One explosion sent a tank soaring high into the air, trailing flames like a huge rocket. It landed near a warehouse.

The blaze was believed to have been caused by a large tank truck which entered the area to load up with vinyl acetate—an ester—bumped into a loading dock, and caught fire. Flames quickly spread to a nearby 30,000-gallon tank, eventually involving twenty similar tanks before being brought under control four hours later.

This was the fire referred to in Chapter 2 as the GATX fire, the first time that AFFF was employed on other than aircraft fires. Its rapid extinguishment by means of AFFF was even more spectacular than the fire.

The aldehydes also contain a carbonyl group. The name *aldehyde* means *al*cohol *dehyd*rogenated because an aldehyde can be produced by subtracting a couple of hydrogen atoms from an alcohol. Notice the difference between ethyl alcohol and acetaldehyde in Figure 3–15.

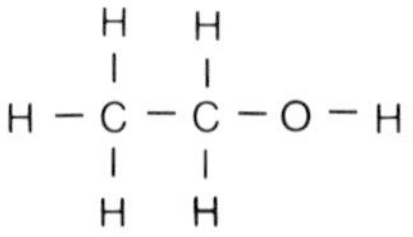

```
    H   H
    |   |
H - C - C - O - H
    |   |
    H   H
```

Ethyl Alcohol
C_2H_5OH

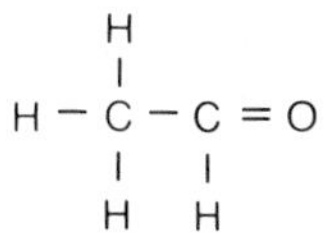

```
    H
    |
H - C - C = O
    |   |
    H   H
```

Acetaldehyde
CH_3CHO

FIGURE 3–15. An Alcohol and an Aldehyde

Aldehydes have an enormous variety of industrial uses, from leather tanning to imparting wet strength to paper, from sedatives to embalming, from plastics manufacture to food flavoring. Two of their number, acetaldehyde and formaldehyde, are of particular interest.

Acetaldehyde

Acetaldehyde is a colorless liquid with a sickening fruity odor. Its boiling point is extremely low, 69°F. (21°C.). It will vaporize completely in a normally heated room. Acetaldehyde will react, often violently, with hydrogen cyanide, hydrogen sulfide, anhydrous ammonia, oxidizing agents, alcohols, ketones, caustic alkalis, acid anhydrides, phenols, halogens, amines, and others. Quite a list! It must be isolated from all of these compounds and elements in storage. On top of this, acetaldehyde can react with itself, forming explosive and unstable peroxides or by polymerizing (forming long chain molecules). Acetaldehyde is an eye, skin, and respiratory irritant and is capable of producing severe eye burns. Prolonged inhalation of acetaldehyde vapors acts as a drug.

There is one bright spot in an otherwise dark picture: Acetaldehyde is totally soluble in water. Flooding dilutes it to a point where it will not support combustion.

Formaldehyde

Formaldehyde is a colorless gas with an unforgettable, highly irritating odor. We include this gas here because it is often found in a water solution containing from 37 percent to 55 percent formaldehyde, stabilized against self-inflicted reactions by the addition of various percentages of methyl alcohol (usually 15 percent). The flash point of this solution, called formalin, is 122°F. (50°C.).

Formaldehyde is an eye, skin, and respiratory irritant. The presence of 650 parts per million in the air is an immediate hazard to life. Fires in formalin solutions yield to flooding with water. Table 3–8 (pages 76–77) summarizes the hazards of several aldehydes.

TABLE 3–8. PROPERTIES OF SOME ALDEHYDES, KETONES, AND ESTERS

	FLASH POINT	IGNITION TEMP.	LOWER FLAMMABLE LIMIT(% BY VOL. IN AIR)	UPPER FLAMMABLE LIMIT (% BY VOL. IN AIR)	SPECIFIC GRAVITY (WATER = 1.0)	VAPOR DENSITY (AIR = 1.0)	BOILING POINT	WATER SOLUBLE	COMMENTS
Acetaldehyde (Ethanal, Acetic aldehyde)	−36° F. −38°C.	365° F. 185°C.	4.1	55.0	0.8	1.5	70° F. 21°C.	Yes	Colorless liquid, unpleasant fruity odor. Irritating and narcotic. Very reactive and potentially unstable.
Benzaldehyde	148° F. 64°C.	377° F. 192°C.			1.1	3.7	355° F. 179°C.	No	Water may blanket fire. Low toxicity.
Butyraldehyde (n) (butanol)	20° F. −7°C.	446° F. 230°C.	2.5		0.8	2.5	169° F. 76°C.	No	Pungent odor. Low inhalation toxicity. May produce eye and skin burns.
(iso)	−40° F. −40°C.	490° F. 254°C.	1.6	10.6	0.8	2.5	142° F. 61°C.	Slightly	
Formaldehyde (gas) (37% solution)	185° F. 85°C.	806° F. 430°C.	7.0	73.0		1.0	−6° F. −21°C.	Yes	Eye, skin, and respiratory irritant. Often found in water solutions.
	122° F. 50°C.						214° F. 101°C.		
Paraldehyde	96° F. 36°C.	460° F. 238°C.	1.3		0.99	4.5	255° F. 124°C.	Slightly	Freezes at 54°F. (12°C.). Sedative with hypnotic effects. Poisonings rare.
Propionaldehyde (proponal)	15° F. −9°C.		3.7	16.1	0.8	2.0	120° F. 49°C.	Slightly	Unpleasant fruity odor. Toxic when inhaled, avoid breathing vapors.
Acetone (Dimethyl ketone)	0° F. −18°C.	1000° F. 538°C.	2.6	12.8	0.8	2.0	134° F. 57°C.	Yes	Low toxicity. Familiar minty odor. UL Classification of 90.
Mesityl oxide	87° F. 31°C.	652° F. 344°C.			0.9	3.4	266° F. 130°C.	Slightly	Toxic when inhaled or by skin contact.

TABLE 3–8. PROPERTIES OF SOME ALDEHYDES, KETONES, AND ESTERS (CONT.)

	FLASH POINT	IGNITION TEMP.	LOWER FLAMMABLE LIMIT (% BY VOL. IN AIR)	UPPER FLAMMABLE LIMIT (% BY VOL. IN AIR)	SPECIFIC GRAVITY (WATER = 1.0)	VAPOR DENSITY (AIR = 1.0)	BOILING POINT	WATER SOLUBLE	COMMENTS
Methyl ethyl ketone (M.E.K.)	21° F. −6°C.	960° F. 516°C.	1.8	10.0	0.8	2.5	176° F. 80°C.	Yes	Smells like acetone. UL rating: 85–90. Somewhat irritating to eyes.
Methyl isobutyl ketone	73° F. 23°C.	860° F. 460°C.	1.4	7.5	0.8	3.4	250° F. 121°C.	Slightly	Vapors are irritating and may become narcotic at high concentrations.
Methyl propyl ketone	45° F. 7°C.	941° F. 505°C.	1.5	8.0	0.8	3.0	216° F. 102°C.	Slightly	Vapors are irritating and may become narcotic at high concentrations.
Amyl acetate (iso)	77° F. 25°C.	713° F. 379°C.	1.0	7.5	0.9	4.5	290° F. 143°C.	Slightly	Banana oil. Somewhat irritating.
Butyl acetate (n)	72° F. 22°C.	790° F. 421°C.	1.7	7.6	0.9	4.0	260° F. 127°C.	Slightly	UL Classification of 50–60. Avoid prolonged contact with skin.
(iso)	64° F. 18°C.	793° F. 423°C.			0.9	4.0	244° F. 118°C.	No	
Ethyl acetate (Acetic ether)	24° F. −4°C.	800° F. 427°C.	2.5	9.0	0.9	3.0	171° F. 77°C.	Slightly	UL Classification of 85–90. Sale is controlled by Government. Not toxic.
Ethyl silicate	125° F. 68°C.				0.9	7.2	334° F. 168°C.		Decomposes in water. Can cause extreme irritation to eyes and nose.
Methyl vinyl acetate	60° F. 16°C.				0.9	3.5	207° F. 97°C.	Slightly	Carries a DOT red label when shipped. Low order of toxicity.
Propyl acetate (n)	58° F. 14°C.	842° F. 450°C.	2.0	8.0	0.9	3.5	215° F. 102°C.	Slightly	Somewhat irritating and toxic. High concentrations of vapor to be avoided.
(iso)	40° F. 4°C.	860° F. 460°C.	1.8	8.0	0.9	3.5	194° F. 90°C.	Slightly	

Esters

The esters are a large group of flammable liquids, often with pleasant odors reminiscent of bananas, pears, pineapples, apples, and other fruit. Esters are used as food flavorings and solvents, in inks and antibiotics, in explosives and vitamins—a striking combination! (See Table 3–8.)

Animal and Vegetable Oils

By now, we have discussed a sizable number of flammable liquids. Some may be unfamiliar, making us wonder why they are so important. Let us emphasize again that the fire potential and toxicity hazards of familiar industrial and household liquids, such as paints, varnishes, lacquers, glues, inks, and paint removers, often depend upon the solvent within them. Most of the liquids discussed in this chapter are used as solvents, and we have only scratched the surface. Hundreds of different solvents are needed to dissolve the multitude of solids industry wants to use and sell in a liquid state. We have left one major constituent of paints, inks, and glues—LINSEED OIL—for the conclusion of this section on liquids. Actually linseed oil is not a solvent, but a drying oil. No paint or ink would be complete without something which will dry and harden upon exposure to air.

Linseed oil has only one real hazard. Its flash point is a safe and sane 403°F. (206°C.) for the boiled type and 432°F. (222°C.) for raw; its boiling point is a little over 600°F. (315.5°C.). But linseed oil does oxidize in air, leading to spontaneous heating. When we discussed spontaneous ignition in Chapter 1, we used a pile of rags soaked in linseed oil as an example. Most flammable liquids are *not* subject to spontaneous heating. Of all those

TABLE 3–9. RELATIVE HAZARDS OF ANIMAL AND VEGETABLE OILS

HIGH HAZARD	MODERATE HAZARD	LOW HAZARD
Cod Liver Oil	Corn Oil Cotton Seed Oil	Black Mustard Oil Castor Oil
Colors in Oil	Olive Oil Paints*	Coconut Oil Lanolin
Fish Oil	Pine Oil	Lard Oil
Linseed Oil	Red Oil Soybean Oil	Oleic Acid Oleo Oil
Menhaden Oil	Tung Oil Whale Oil	Palm Oil Peanut Oil
Perilla Oil		Turpentine

*Paints which contain a dry oil.

surveyed in this chapter only one has this tendency—turpentine. Petroleum products are not guilty, although often accused. Yet, many fire officers have a mistaken idea about this fire cause. Before marking down "spontaneous ignition" as the probable cause in a fire report, consider whether the suspected liquid came directly from a LIVING SOURCE. If the fire did not involve one of the liquids listed in Table 3–9, investigate its "cause" further.

QUESTIONS ON CHAPTER 3

1. Gasoline has a flash point about:
   a. −65.2°F. (−54°C.)
   b. −45°F. (−43°C.)
   c. +32°F. (0°C.)
   d. +300 to 400°F. (149 to 204°C.)

2. The boiling point of ethyl amine is:
   a. −40°F. (−40°C.)
   b. 0°F. (−18°C.)
   c. +62°F. (17°C.)
   d. +723°F. (384°C.)

3. The ignition temperature of ethyl ether is:
   a. −49°F. (−45°C.)
   b. +95°F. (35°C.)
   c. +356°F. (180°C.)
   d. +1000°F. (538°C.)

4. Most ethers have a tendency to foam explosive peroxides. This is particularly true in the case of:
   a. Isopropyl ether
   b. Ethyl ether
   c. Butyl ether
   d. Divinyl ether

5. Forming emulsions with a coarse spray can be an effective method of extinguishing fiıes in:
   a. Kerosene
   b. Alcohols and ketones
   c. Aldehydes
   d. Heavy oils

6. One of the problems of naphtha is:
   a. High toxicity
   b. It will explode spontaneously
   c. Wide flammable range
   d. Several different kinds, with different flash points.

7. The most toxic common hydrocarbon is:
   a. Gasoline
   b. Benzene
   c. Turpentine
   d. Hexane

8. Which of these liquids is *not* subject to spontaneous heating:
   a. Linseed Oil
   b. Gasoline
   c. Turpentine
   d. Cottonseed Oil

BIBLIOGRAPHY

National Association of Mutual Casualty Companies:

Technical Guide 6, "Handbook of Organic Industrial Solvents"

United Air Lines:

"Aircraft Familiarization for Firemen and Rescue Crews." *The American Insurance Association* (formerly the National Board of Fire Underwriters).

American Insurance Association:

The AIA has produced a series of reports and bulletins on the flammable liquid problem.

Research Report 5, "Processes, Hazards, and Protection involved in the Manufacture of Spirituous Liquors"

Special Interest Bulletins:

No. 7 (July 3, 1934), "A Tragic Fire and Its Lessons"

No. 161 (January 1957), "Physical and Chemical Properties of Flammable Liquids and Gases"

No. 181 (June 1956), "Portable Kerosene Heaters"

No. 182 (January 1958), "Jet Aircraft Fuels"

No. 215 (May 1961), "Paint Coatings—Fire Hazards"

No. 288 (revised October 30, 1950), "Paints and Cleaning Materials Fire Hazards"

Union Carbide Company:

Has a series of beautifully produced pamphlets on flammable liquid properties. Titles include: "Esters," "Ketones," "Ethylene Amines," "Alkyl Amines," "Ethanol," and many more.

4 Flammable Liquids in Bulk

Construction requirements for flammable liquid tanks, pipings, valves and fittings, their installation and location, is an enormous subject. The storage code for flammable liquids forms a major part of any fire prevention course. We will give particular attention to the problems fire fighters encounter when fires and explosions occur in bulk flammable liquid storage, in tanks, tank trucks, warehouses, and underground pipelines.

Why do flammable liquids figure so prominently in annual reviews of the biggest fires? Not only because of their inherent hazard, but also because of the enormous quantities involved in storage and shipping. A single large tank may contain 200,000 barrels of fluid or more. A tank farm or refinery may have several dozen such tanks, each a potential headline. Flammable liquid drum storage may be concentrated in a yard or warehouse, with one 55-gallon drum piled atop another and hundreds in a small area. A tank truck can carry thousands of gallons of gasoline along a busy street for delivery to the service station on the corner.

Fortunately, bulk flammable liquid fires are rare, and a fire department is seldom called upon to face such an emergency. The best way to prepare for such incidents is through examination of the victories and defeats of other fire fighters. One of the most valuable services provided by the NFPA and the AIA is their thorough investigation and reporting of these large-loss fires. These reports deserve careful study. (See the bibliography at the end of this chapter.)

Tank Fires

Much of this section states the obvious, but overlooking the obvious is all too common on the scene of a large flammable liquid fire. The sizeup of such a fire can be extraordinarily difficult. Radiated heat may make close approach impossible, smoke may obscure vision. Yet the fire fighter must evaluate a great many factors correctly when facing a tank fire.

The Water Situation

Hopefully, you will already know the location of useful hydrants and the size of the mains which feed them. In several bulk fire histories, hydrants were in poor position, requiring long, awkward, time-consuming leads. In other cases, water supplies were insufficient. Make no mistake about it, the amount of water available is vital to your operations. A great deal of heat may have to be dissipated before fire fighters can approach within working distances. Only the use of protective water curtains may allow them to close necessary valves. Cooling water sprays *must* be applied to maintain the integrity of exposed tanks. In some cases, the volume of water used for cooling has literally caused flammable liquid tanks to float. Guard against this by ensuring water drainage. If there is not enough water to use for protecting exposures, the evacuation of fire fighters to safe distances, at least 2000 feet (600 meters), is required!

Liquids Involved

This is where you put your knowledge of flammable liquids to work. What is their specific gravity or water solubility? It may be possible to blanket, emulsify, or dilute them. If their flash point is in the right range, direct application of a fog pattern may bring quick extinguishment. Your knowledge of flammable liquids will also tell you what to expect from them. You will know the proper type of foam to use. You won't waste your time or manpower in a futile effort to extinguish a large gasoline fire with water. You will be aware of the possibility of steam explosions and boilovers.

We must pause here to examine one type of boilover in more detail. CRUDE OIL, as you will remember, is a mixture of petroleum fractions. Before it is refined, crude often contains a great deal of water, has a wide range of boiling points, and is extremely viscous. In a storage tank, the water eventually separates and goes to the bottom. After a crude oil tank is on fire for several hours, the lighter fractions have burned off. This leaves hot, heavier fractions which sink toward the bottom of the tank in what is called a "heat wave." We can observe the downward progress of a heat wave by the use of a strip of heat-detecting paint or by watching the action of water flowing down the sides of the tank; water is likely to vaporize when it reaches the area of the heat wave. Heat waves travel at a varying rate, approximately one to four feet per hour. When they reach a point five feet (1½ meters) above the known level of the bottom water, it is time to leave the scene. When steam is formed, it is trapped by the heavy viscous crude oil, and the expanded contents of the tank are violently expelled. A crude oil boilover is signaled by an increase in flame intensity. It is highly unlikely that dikes will contain the boilover unless they are especially constructed for the purpose. At one bulk crude oil fire, many fire fighters were hurt when a massive boilover of hot liquid took place. Most of them thought the tank had exploded. Finally, don't be in a hurry to return; successive boilovers in crude oil tanks are not uncommon.

The Physical Layout

Preplanning will develop a detailed picture of the physical layout. Are there dikes to confine a spill fire to a localized area? If so, only one, or relatively few tanks, will be immediately involved. Since preventing spread is the primary consideration in any flammable liquid fire, dikes solve the problem of confinement, although another problem immediately takes its place: Each of the tanks in the involved area will be subjected to a great deal of stress as fires continue to burn around them. Enough water to protect the tanks can overflow the dikes, destroying their value. Can these burning liquids be safely drained away into impounding basins where, away from exposures, they can burn out? Only the value of the liquid, often the cheapest part of the potential property loss, will then be lost.

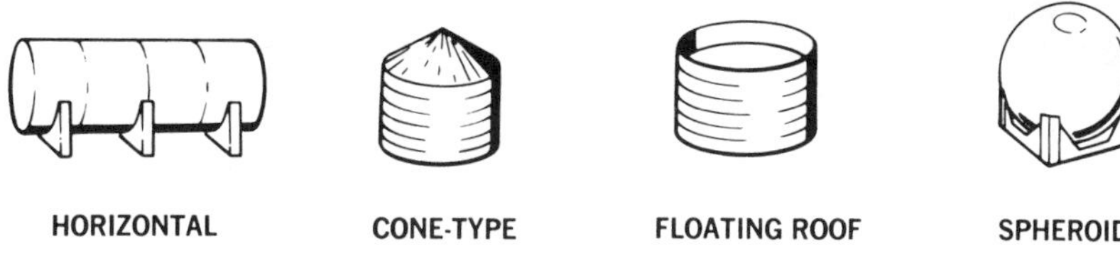

FIGURE 4–1. Examples of Tank Design

Types of Tanks and Piping

Preplanning can also tell you the types of tanks in the occupancy and the location of piping, vital bits of information. As Battalion Chief Erven of the Los Angeles Fire Department has commented, "When fighting tank fires, often the only method of extinguishment is to pump out the content of the tanks. This is why modern tank farms place valves outside the dikes. In one large tank fire at Union Oil in Wilmington, thousands of barrels of gasoline from the burning tanks were pumped into empty tanks 15 miles away" (see Figure 4–2, page 84). Most tanks have vents designed to prevent vapor overpressures.* You should know if these vents are merely sufficient for the normal "breathing" of the tank. Inadequate vents, such as these, are incapable of

*Spheroidal and cone-type tanks also have vents for relief of overpressures.

relieving the tremendous increase in vapor pressure caused by a surrounding spill fire. Tanks, vented in this way, may explode like overheated boilers, raining down thousands of gallons of burning liquid. The increasing sound of vapors escaping from the vents will indicate the buildup of pressure in an overheated tank. This sound has been variously described as "a whistle," "a jet engine," and so forth. Unless water can be immediately applied to such a tank to cool it and lower the interior pressure, the area should be

FIGURE 4–2. Gasoline Tank Farm Fire (Photo by Dale Magee)

vacated. The first fire-fighting efforts should be devoted to extinguishing any ground fire which is heating the tank.

When there is a fire at the vent, you must maintain a cooling water spray over the area involved to prevent localized overheating, sometimes followed by a violent rupture of the entire tank. This is what happened to the pentane-hexane spheroid discussed on pp. 46–49. But there is an exception to this rule, pointed out by Miles Woodworth, NFPA Flammable Liquids Engineer, in an article in *Fireman's Magazine,* July 1956.

> A fire that is burning at a tank vent with a yellow-orange flame, emitting black smoke, indicates a vapor-rich condition within the tank, which is above the flammable or explosive limits. This type can be extinguished by smothering with steam, dry powder, CO_2, or wet blankets. There is no danger of an explosion.
>
> Fire at a tank vent that burns with a snapping, blue-red, nearly smokeless flame indicates a vapor-air mixture within the tank that may be explosive. There is constant danger of an explosion should the flame reach the inside of the tank. The best attack procedure, therefore, is to maintain a "positive pressure" within the tank by pumping liquid into it. When a vapor-rich condition is indicated by change in the flame character to a smoky yellow-orange flame, danger of explosion has passed. Extinguishment can then be accomplished as stated in the previous paragraph.
>
> Never apply cooling water to tanks burning with a snapping, blue-red, nearly smokeless flame at the vent. Nor should such tanks be pumped out. Several of the tank vent fires reported to the NFPA featured injury to firemen when water applied to cool the tank shell caused an explosion.

As you can readily see, this can be a situation where you are damned if you do, damned if you don't. Sometimes, there is just no completely acceptable solution to the problems posed by a bulk flammable liquid fire.

BLEVEs

Boiling Liquid Expanding Vapor Explosions—BLEVEs—can occur whenever flammable liquids are rapidly heated to temperatures well above their boiling points.

The resulting expansion of the vapors causes internal pressures beyond the tank venting mechanism capabilities, followed by the inability of the tank structure itself to withstand the high internal pressures. This combination of factors can result in an instantaneous release and ignition of the flammable vapors caused by structural failure of the tank. The explosion, called a BLEVE, then occurs. There is often a time delay prior to the explosion, a factor that can be fatal, for the time delay allows both fire-fighting personnel and equipment, as well as spectators, to approach close enough to become potential casualties of the extremely violent explosion. A BLEVE is usually accompanied by a huge fireball.

A BLEVE is not limited to stationary storage tanks, but can involve railroad cars and tank trucks as well. It can be destructive to life and property over a radius of more than 2000 feet (600 meters).

The approach to suppression, or in these instances *prevention,* of a BLEVE is to immediately place very heavy stream appliances into operation. They should be positioned where the water can be played on the upper sides and top of the tanks involved. If at all possible, these heavy streams should have at least a 500-gallon-per-minute (2000-liter-per-minute) capacity and should be unmanned.

Horizontal Tanks These larger versions of the familiar railroad tank car have their own peculiar way of reacting to excessive pressures. They may explode like other tanks, but more often they skyrocket. One end of the tank gives way and the tank rockets off its supports in response to the same principle that sends a missile into orbit from its launching pad at Cape Kennedy. A fundamental rule in tank fire fighting is *never approach an involved horizontal tank from either end.* Always approach it from the side.

Unfortunately, this rule was not so clearly established before a fire in the Midwest a decade or so ago. *After* the fire, of course, everybody knew the rule. In any event, fire fighters were stationed at the ends of several horizontal tanks, fighting a severe flammable liquid spill fire. Intent upon confining the fire to a small area, they used hose lines to sweep the burning gasoline back under the tanks instead of flushing it into an open area. The vents on this group of tanks were too small to relieve the tremendous overpressures that resulted. As a result, one tank rocketed directly at a group of firemen ninety feet away, killing six of them. (Horizontal tanks can fail still another way. If they sit upon unprotected steel supports, instead of concrete or masonry, these supports can buckle quickly when involved in a fire. The tank hits the ground or the piping breaks. Either way, the contents are released and an explosion often results.)

Floating Roof Tanks As its name implies, this type of tank eliminates potentially explosive vapor space by having the roof rest directly upon the liquid contents. Instead of vents, this type of tank is built with a deliberately weakened roof seam. Flammable liquid fires often begin with an explosion, and a tank with its roof ruptured or missing is not subject to a pressure explosion. But when fire fighters reach it, it will probably be on fire, and heat will cause the sides of the tank to fold in a few feet above the level of the liquid.

The best way to extinguish an OPEN TANK FIRE is through the application of the proper foam blanket. *No extinguishing agent is as successful as foam on large flammable liquid tank fires.* When applying cooling streams of water on a tank equipped with a floating roof that is exposed by an adjacent fire, fire fighters must be careful not to exceed the capacity of the scuppers. At a marine oil terminal fire, a fireboat applied such a large quantity of water to prevent a 60,000-barrel tank of gasoline from being involved that the tank top became overloaded and sank to the bottom. This resulted in a large open tank of gasoline that had to be protected from ignition by the maintenance of a thick blanket of foam.

Perhaps the only fairly safe flammable liquid tanks are those two feet underground. Such tanks are almost impervious to a fire above them. If there are no exposed or exploding tanks to hold your attention, you should give consideration to the piping. Involved pipes may fail and release the entire contents of a tank into a running spill fire, thus creating a fire that is extremely difficult to combat. A foam blanket underneath vulnerable piping may help control the situation until the proper valves can shut off the flow.

Plant Manager

Get the plant manager's advice if you can. He can be of invaluable assistance if he is on the premises and coherent. He can tell you the location of valves that will shut off the flow from broken piping. He knows what types of liquids are involved. He can point out the location of dike drains and where they lead. He can estimate the level of the bottom water in a crude oil tank. He can assist you in getting foam-generating equipment. He can show you possible ignition sources and how to shut them off if you have a flammable liquid spill not yet on fire. Unfortunately, it is possible for him to be many miles away—on vacation. Do not be totally dependent upon help from this source. However, there is usually at least one knowledgeable person on duty. Solicit suggestions. If this person does not have the necessary information, he or she will know where to obtain it.

Keep the Public Away!

Have the police department keep spectators and all persons other than fire fighters back at least 2000 feet (600 meters). This is most important. If a large tank lets go, the amount of heat released is overwhelming. You have enough of a problem protecting fire fighters from this catastrophe without worrying about the public. Although this may seem to be unneeded advice, the history of bulk flammable liquid fires has proven their crowd appeal and the danger to onlookers. Keep them away even though it will spoil their photographs.

Drum Storage

Let's face a few facts about flammable liquid drum storage. Rules and recommendations for safe handling and storage are more often ignored than followed. Companies tend to use every possible inch of storage space. As a result, the number of drums in a storage area will exceed requirements for safe size and height of piles, and for the permissible quantity within a particular area. The maximum number of drums allowed in a particular sprinklered area depends upon the location of the storage space, whether it is detached, cut off, or part of the main building. The type of liquids stored is also a factor. (For an excellent summary of this entire question, see Chapter 47, "Handbook of Industrial Loss Prevention," Factory Mutual Association 1968.)

Flammable liquid storage is not always strictly segregated or in detached, noncombustible buildings with dispensing done in cutoff, properly arranged, and protected rooms. Ignition sources can be present in both areas. Housekeeping may be substandard with no inspection program for leakage. Drainage systems can be inadequate; sprinkler systems, nonexistent. Most of the fire fighters with experience in industrial inspection know that all companies do not strictly observe safety practices. Some do; others are very haphazard and negligent. It can be a long time between fire department inspections. This is another reason why flammable liquids cause an appreciable percentage of industrial fires.

In an unopened drum of good construction, a flammable liquid is only a potential hazard. Nothing will happen as long as it is safely tucked away. Trouble begins when the liquid gets loose in some way: by leakage caused by rough handling or falling from a high pile; by deterioration of the container; by unsafe dispensing practices; by structural failure of the drum because of heat or a nearby explosion. The amount of trouble, as we have seen, will depend on the type of liquid set free, potential ignition sources, and the kind of protective equipment the occupancy affords.

Once again, the particular properties of a given liquid (see Chapters 1–3) determine its degree of hazard. When a liquid is liberated, for whatever reason, it must do what its properties dictate. Vapors will travel according to their densities, and liquids will seek their level. If a vapor cloud within the flammable range encounters an ignition source above the ignition point, an explosion *always* occurs.

Companies that store drums of flammable liquids can reduce the risk of fire and explosion in a number of ways. Good ventilation of rooms where flammable liquids are stored or handled can be vital. Ignition sources can be eliminated wherever possible: open flames, sparks, electrical equipment, static, hot surfaces, radiant heat, friction, overheating, spontaneous ignition, and all other sources that might not be considered until it is too late.

But, even when the best practices are followed, including safety precautions, trouble can and does start. When it does, the most efficient safeguard in areas where flammable liquid drums are kept is the automatic sprinkler system. Of course, sprinklers have limitations. The amount of drum storage in a room can exceed sprinkler control capability. Explosions can damage the system, knocking it out. A small leak from a drum can provide fuel for a localized fire intense enough to rupture nearby drums, but not big or hot enough to activate the sprinkler system. The flash from one bursting drum can open hundreds of sprinklers, some well beyond the initial fire area, and severely overtax the system. Finally, sprinklers will not extinguish many low-flash-point flammable liquids, and flowing water may carry burning liquid, spreading the fire to uninvolved areas. Nevertheless, a sprinkler system can prevent a disaster. It will cool surrounding drums and exposures. With proper drainage, it can flush away the burning contents of a drum before exposed drums are heated dangerously. Without sprinklers, the heat of a spill fire can cause a pile of drums to explode like a pan of popcorn.

Upon arrival, fire fighters complete the work sprinklers have begun. A fire fighter has one big advantage over a sprinkler system: he or she can think. The fire fighter makes sure the system is not turned off prematurely, supports it through fire department connections, then sets about confining the fire and using the proper extinguishing agents.

Tank Trucks

A truck driver is not a fire fighter and should not be expected to know the contents of tanks unless he or she is delivering gasoline. Frequently, the driver neither knows nor cares, but he or she does have bills of lading which will tell us exactly what liquid is in the tanks. A teamster can be of great value to fire fighters because such a person usually knows the operation of the truck and is familiar with the location of valves.

All fire fighters should take a good look at a tank truck, if they have not already done so. They are beautifully designed to reduce the possibility of a fire. Internal valves will close to prevent the drainage of the tank if piping breaks. These valves have remote controls. If the vents on a modern tank truck are free to operate (and they will, if the truck is not overturned) they will take care of overpressures. There has been no record of a tank truck rupturing or exploding for this reason. Any explosions which occur come from the tires, the truck's own fuel tanks, or sudden ignition of spillage.

Some tanks are made of steel and some of aluminum. Steel tanks, of course, are stronger and less likely to break open if the truck overturns. Aluminum tanks will melt far more rapidly if the tank is involved in a spill fire. This is not as dangerous as it sounds. It presents us with the relatively simple problem of several small, open tank fires as the aluminum melts down just above the liquid level to reveal the internal compartments of the tank. These small fires can quickly be extinguished by foam or by dry chemical, supported by fog streams for cooling and prevention of re-ignition.

If the truck is overturned, running-liquid spill fires can be controlled by hose streams and directed to safe areas for burnout. A spill that is *not* on fire is more complicated. Depending on the severity of the spill, it may be necessary to shut off ignition sources or even evacuate downwind areas. Foam blankets can be employed. The spilled liquid may be able to be flushed down storm sewers, although this practice is frowned upon. Vacuum tank trucks, such as those used in the oil fields or those utilized to pump out septic tanks, will quickly and effectively suck up flammable liquid spills and pools. The following procedure is recommended in NFPA Bulletin 328M.

> Before washing spilled petroleum products from street surfaces into drains or sewers, which may be a potentially dangerous action and often is an unlawful practice, other disposal means should be considered such as mopping up with sand, rags, or mops. If, in an emergency, no alternative is available, disposal into a drain or sewer should be done only on the decision of a qualified person, after the chief of the fire department has been notified.
>
> Disastrous consequences may result from the thoughtless or deliberate dumping into drains of waste products which are either directly flammable or which, by reaction with organic matter in sewers, may produce a flammable mixture.
>
> [When] volatile flammable liquids accidentally enter a drainage system because of a spill or other emergency, steps should be taken to minimize the hazard by exhausting the vapors with blowers or exhaust fans driven by explosion-proof motors and by pumping out the liquid with pumps equipped with explosion-proof motors. Floor-drain openings in the area of the spill and for some distance downstream should be checked for escape of vapors. Water should be placed in any dry traps to seal them. Copious quantities of water should be used to flush any flammable liquid through the system quickly and to dilute it, if miscible with water.*

Service Stations

Gasoline tank trucks are generally heading for a neighborhood service station. With the exception of car fires, this type of occupancy constitutes our greatest involvement with flammable liquids. There are over 200,000 service stations in the United States. Each of them sits above at least one gasoline tank. Although some of the following safety precautions are impossible to enforce, they are directed at the causes of most service station fires.

---

*NFPA 328M, "Flammable Liquids and Gases in Manholes and Sewers."

1. Do not permit smoking when gasoline is being dispensed. Post clear signs to that effect.

2. Motorists should turn off their ignition when their tanks are being filled.

3. A competent attendant should be in the vicinity of the vehicle being filled, even though automatic nozzles are being used.

4. Above-ground gasoline tanks are prohibited.

5. Sale of gasoline in glass jugs or bottles is forbidden.

6. When tank trucks are making a delivery, they should be on service station property, not on the street.

7. Waste flammable liquids must not be dumped into sewers.

8. Open flames are forbidden.

From a fire-fighting standpoint, the most important requirement for service stations is a clearly identified switch, readily accessible, that cuts off electric power to the pumps in the event of a fire or physical damage to the dispensing units. Our strategy should include the immediate closing of this switch to prevent a malfunction of the pumps, causing an increase in the amount of gasoline involved. Beyond this, the primary consideration remains the same: confine the problem. Protect exposures, don't wash gasoline toward them. Let the fire burn out if it cannot be quickly extinguished by portable foam or dry chemical extinguishers. If there is an extensive spill fire, with resulting danger to exposures, sterner measures must be employed, such as coordinated 1½″ fog lines. Directed with intelligence, fog lines can surround and direct the course of the spill, even though they will not extinguish it.

High-Pressure Pipelines

A street construction crew was digging through the center divider of Venice Boulevard near the West Los Angeles–Culver City limits on the morning of June 16, 1976. The claw of a trencher tore into a high-pressure gasoline pipeline, causing a rupture that spewed the gasoline high into the air and across three traffic lanes. Seconds later, an explosion rocked a wide area of the two cities (see Figure 4–3, page 92).

The Culver City and the Los Angeles city fire departments made an immediate response into the area with greater than normal first alarm assignments because of the reported explosion. As the first-due companies arrived, they found a row of buildings, a city block in length, fully involved

FIGURE 4–3. Fire from High-Pressure Gasoline Pipeline Rupture (Photo by Phil McBride)

in fire. People who had been inside the stores, the structures, and the vehicles on the street were caught in the gasoline-fed holocaust.

Many were badly burned. Walls blown in by the blast collapsed, causing numerous other injuries to occupants. One woman suffered a heart attack and died. Another was trapped in her car and was burned to death. Others died later from burns.

Fire fighters applied large quantities of AFFF, which helped to confine the gasoline blaze until the pipeline could be drained. Automatic valving shut down the line system seconds after the rupture, but the residual fuel drained and burned for about 25 minutes. The fire was contained about 14 minutes later. An estimated 16,000 gallons of gasoline had been ignited in this tragic accident.

High-pressure pipelines carrying a great variety of flammable liquids (as well as toxic and flammable gases) pass beneath city streets today. Do you know where they are located? Do you know what they transport? Do you have a plan of action in the event of an accident?

Even an eight-inch pipeline can provide you with a flammable liquid—*in bulk*.

QUESTIONS ON CHAPTER 4

1. When evacuation of fire-fighting personnel at an oil tank fire is required, the *minimum* safe distance is:
   a. 100 feet (30 meters)
   b. Beyond the smoke
   c. 2000 feet (600 meters)
2. A crude oil "boilover" is signaled by:
   a. An increase in black smoke
   b. Jets of flaming oil
   c. An increase in flame intensity
3. The primary consideration in controlling a flammable liquid fire is:
   a. Using fog
   b. Immediate ventilating
   c. Preventing spread
   d. Toxic vapors
4. A snapping, red-blue, nearly smokeless flame, burning at an oil-tank vent, shows us that there is
   a. No danger of explosion
   b. A vapor-rich condition in the tank
   c. Great danger of explosion
5. Never approach a horizontal flammable liquid tank that is on fire:
   a. From the sides
   b. Without using fog
   c. Without contacting the plant manager
   d. From the ends

VISUAL AIDS

"BLEVE," a film produced in January 1976 by the NFPA, has been a "best-seller." The October 1975 issue of *LP-Gas* devoted a number of pages to a review of the film: "It's a film of stark reality which may *offend* some LP gas dealers . . . but it's also a film every dealer in the country should see.'' So should every fire fighter.

Two excellent color movies, produced by the National Fire Protection Association, are: ''Fighting Tank Fires,'' 25 minutes; ''Tank Vehicle Fire Fighting,'' 30 minutes.

Two interesting movies about actual fires are: ''Analysis of a Bulk Plant Fire'' (the Kansas City tragedy), available from Standard Oil Company; ''Refinery Fire at Signal Hill.'' Each film runs about twenty-five minutes, available from the Los Angeles Fire Department.

BIBLIOGRAPHY

Fireman's Magazine, published by the NFPA, has had a series of articles on the problems of bulk storage of flammable liquids. Some of these include:

''This Year's Biggest Industrial Fire'' (October 1955).

''Flammable Liquids Notebook'' (February and April 1956).

''Texas Refinery Tragedy'' (September 1956).

''Memorize These Tank Truck Design Features'' (August 1959).

''Tank Blast Kills Six, Injures Sixty-Four!'' (November 1959).

''Horizontal Tank Fires—How to Handle Them'' (August 1960).

''What to Do about Gasoline Service Stations'' (February 1961).

The American Insurance Association (formerly the National Board of Fire Underwriters) has produced a series of reports and bulletins on the flammable liquid problem:

No. 40 (August 1, 1937), ''Oil Fire, Atlantic City, New Jersey.''

No. 70 (January 17, 1939), ''Cities Service Oil Co. Fire.''

No. 110 (revised November 1957), ''Flammable Liquid Storage and Handling in Drums and Other Containers.''

No. 128 (September 5, 1941), ''Oil Tank.''

Fire Command (Firemen before July 1970), published by the NFPA, has had several articles involving BLEVEs:

''Lessons from an LP-Gas Utility Plant Explosion and Fire'' (April 1972).

''Hazardous Materials Transportation Accidents'' (April 1974).

''The Terrible Blast of a 'BLEVE' '' (May 1974).

''Seven Fire Fighters Injured in LP-Gas 'BLEVE' '' (August 1974).

''Coordinated Attack Limits Post-Blast Damage'' (July 1975).

''Two Train BLEVEs: Different Situations Require Different Strategies'' (September 1976).

''Three Fire Fighters Die When Exposure Hazard Ignored—An NFPA Fire Study'' (January 1977).

5 Pressurized Gases

Solids retain their size and shape. Liquids have a definite size and at least one shape, a flat top surface. But gases have neither—no size, no shape, and, seemingly, no rhyme nor reason.

General Principles

Gases will expand upon heating (indefinitely, if you let them) and contract upon cooling. They are completely elastic and will fill any container. Furthermore, they will not settle in it. If you put one liter of liquid in a two-liter container, one liter of liquid will flow to the bottom; put one quart of gas in a two-quart container and you have two quarts of gas. The gas pressure will be the same at the top as it is at the bottom.

Different gases will mix completely. If you put a lightweight gas and a heavyweight gas in the same container, there will not be the sharp line of division you would find in a bottle of mixed gasoline and water. Even gases with different vapor densities will not separate completely. A small amount of the lightweight gas will be found at the bottom of the container, some of the heavy gas at the top.

The only logical explanation for the behavior of gases is that they are composed of tiny particles in constant random motion. The higher the temperature, the more violent the motion, the greater the urge to expand, to fly off in all directions. A great many gas "laws" have been suggested to predict this behavior of gases. Many of them, however, are not easily applied on the fire grounds.*

*Avogadro's Law says that equal volumes of *any* gas contain the same number of molecules. Henry's Law and Roualt's Law have to do with solubility of gases. Dalton's Law predicts the total pressure of a mixture of gases. Graham's Law is a description of the rates of diffusion of gases. If interested, consult a physics textbook for more information.

This is the DOT label (green with black lettering) required on cylinders when compressed, nonflammable gases are being shipped.

FIGURE 5–1. DOT Nonflammable Gas Labels

However, it is valuable to know that the VOLUME of a gas (generally the size of the container in which a gas is stored), the TEMPERATURE of a gas, its PRESSURE, and sometimes its AMOUNT are all interrelated. Change the value of one of these factors and the value of another or perhaps all of them are changed.

Gas Pressure

As fire fighters, we are interested in the pressure of a gas. This is where its danger might lie. How can the pressure of a gas be increased? First, let us repeat that pressure is caused by the impact of gas molecules against the sides of a container that prevents them from heading for parts unknown. The more impact in a given amount of time, the higher the pressure. When we ask how the pressure of a gas can be increased, we are asking how the number of impacts can be increased. There are *three ways,* by changing each of the other factors we mentioned.

The nonflammable gas placard must be green with white symbol, inscription, and ½-inch (12.7-mm) border.

FIGURE 5–1. DOT Nonflammable Gas Labels (cont.)

1. *We can decrease the volume.* A smaller container will shorten the distance a gas molecule has to travel before it contacts the sides. This will increase the number of collisions within a given time period—therefore, more pressure. The amount of pressure is predicted exactly by Boyle's Law. Cutting the volume in half, when the temperature and the amount of gas are held constant, will double the pressure. (Boyle's Law reverses this wording: "at a given temperature, the volume occupied by a gas is inversely proportional to the pressure," but means the same thing.) (See Figure 5–2, page 99.)

2. *We can increase the amount of gas within the cylinder.* More gas means more molecules to collide against the cylinder walls.

3. *We can heat the gas.* A hot gas has faster-moving molecules. They will strike the sides of the cylinder more often than slower-moving molecules.

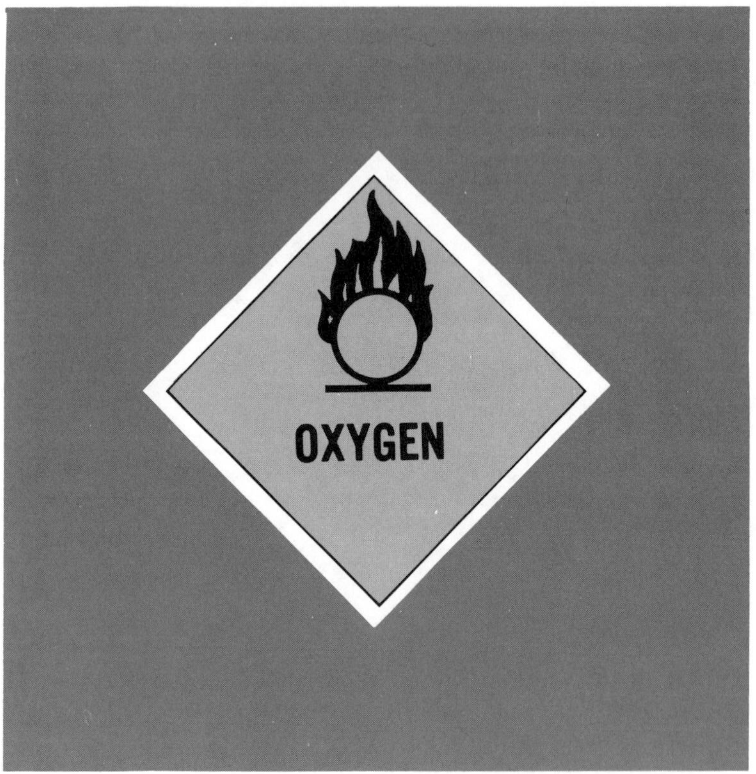

The oxygen placard must be yellow with a ½-inch (12.7-mm) white border. The symbol and inscription must be black.

The chlorine placard, not illustrated, must be a white 10¾-inch (273.0-mm) square-on-point with a ⅛-inch (3.2-mm) black solid line border ½ inch (12.7 mm) in from each edge. The symbol, a skull and crossbones, and the chlorine inscription must be black. The poison gas placard is identical to the chlorine placard in shape, size, symbol, and colors—except that the inscription reads poison gas.

FIGURE 5–1. DOT Nonflammable Gas Labels (cont.)

Combined Gas Law

The amount of pressure increase coming from a rise in temperature can be predicted by the Combined Gas Law (a combination of Boyle's and Charles's Laws). This law can be extremely useful. Most gas cylinders will release their contents through pressure relief devices when the pressures within them reach such a level that the cylinder is in danger of bursting. Whether the released gas is flammable or toxic can be an important factor in

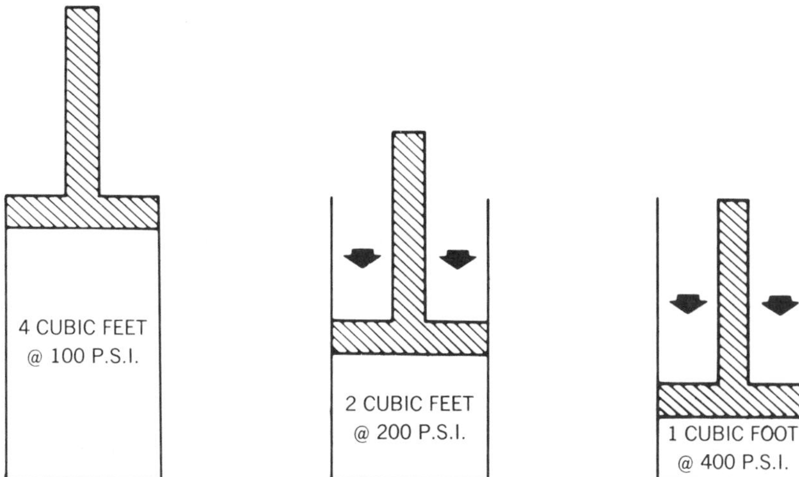

FIGURE 5–2. Volume-Pressure Relationship

a fire situation. An approaching fire can quickly heat a cylinder to 500, 1000, or even 1500 degrees Fahrenheit.

The Combined Gas Law says, in effect, that the pressure of a gas is dependent upon the temperature to which it is exposed and the volume of the container it is in. Since the volume of a steel cylinder expands very little, we can ignore this part of the problem. We are interested in the interrelationship of temperature and pressure. By temperature, we mean absolute temperature: the thermometer reading plus 459.7°F. (273°C.). (Absolute zero is −459.7°F. (−273°C.), often given as −460°F.) By pressure, we mean absolute pressure, which includes atmospheric pressure.

Let us make the arithmetic easy. What would happen to the interior pressure of a gas cylinder at 1000 psia at 40°F. if it were heated to 1540°F.? 40 plus 460 is 500 degrees absolute, 1540 plus 460 is 2000 degrees absolute.

Minus the constant volume factor, our simplified equation looks like this:

$$\frac{\text{Temperature 1}}{\text{Temperature 2}} \text{ is in ratio to } \frac{\text{Pressure 1}}{\text{Pressure 2}} \text{ or } \frac{500}{2000} : \frac{1000}{????}$$

The obvious answer is 4000 pounds pressure within the cylinder, enough to burst it. Many equations would be more difficult, but no more difficult than some problems in hydraulics. There is even a quick rule of thumb we can use: *Beginning with a gas at normal temperatures, an increase of 500 degrees will just about double its pressure; an increase of 1000 degrees will triple its pressure, and 1500 degrees will quadruple it.* In Celsius (metric) measurement, an increase of 273°C. will double the pressure.

Raising the temperature of a gas raises its pressure; lowering the temperature lessens the pressure. Once again, we can reverse this. Raising and lowering the pressure of a gas raises and lowers its temperature. Most of us are familiar with this phenomenon from filling our oxygen or compressed air bottles. As we shall see later, this is the principle that makes refrigeration systems work.

Fire Fighting Pressurized Gases

Industrial gases are shipped and stored in two ways: as liquids which generate a stabilizing vapor pressure, or as gases in high-pressure cylinders of various kinds. Certain hazards, common to all gases stored under high pressure, make it convenient to discuss them as a group. This chapter concentrates on them, although it must be remembered that all arbitrary divisions such as this contain exceptions. Many gases are shipped and stored in both forms.

High-pressure gases are universally valuable: Methane is one of our most important fuels; acetylene and hydrogen are used in welding and cutting operations; oxygen has many medical and industrial uses; ethylene is used both as an anesthetic and to ripen fruit. Gases are used for such diverse purposes as rocket propellants, fumigants, refrigerants, and insecticides. Table 5–1 lists the gases we will examine in this chapter, but this is by no means a complete list of the pressurized gases employed by American industry. Gases with special properties or uses will be part of later chapters, and Chapter 6 will cover liquefied gases, including ammonia and the LP gases.

Several questions must be answered before fire fighting pressurized gas fires can be successful.

Type of Gas Involved

Learning the type of gas involved must be your first objective, for you will be working in the dark until you find out what gas is on fire or loose. Plant personnel may identify a gas, or identification may come from names or formulas on containers and bills of lading. The color and shape of the gas cylinder itself is of doubtful value in identification. Manufacturers tend to use their own color-coding system. These vary. Some manufacturers store all their gases in cylinders of the same color and use them interchangeably. There has been some talk that the Compressed Gas Association will attempt to standardize cylinder colors throughout the industry. This would be of great value. Until this happens, the colors listed in Table 5–2 (page 103) are recommended by the National Bureau of Standards (and sometimes adhered to).

TABLE 5–1. PRESSURIZED INDUSTRIAL GASES

	LOWER FLAMMABLE LIMIT (% BY VOL. IN AIR)	UPPER FLAMMABLE LIMIT (% BY VOL. IN AIR)	IGNITION TEMP.	VAPOR DENSITY (AIR = 1.0)	BOILING POINT	WATER SOLUBLE	TYPICAL CYLINDER PRESSURE (PSI)	COLOR OF REQUIRED DOT LABEL
Acetylene (C_2H_2) (Ethyne; Ethine)	2.5	81.0	571° F. 299° C.	0.91	−119° F. −84° C.	Yes	250	red
Air (Mixture)	(Nonflammable; will support combustion.)			1.0	−317° F. −194° C.		2,000	green
Argon (Ar)	(Nonflammable; will not support combustion.)			1.4	−302° F. −186° C.	Slightly	2,000	green
Carbon dioxide (CO_2) (dry ice)	(Nonflammable; rarely supports combustion.)			1.5	−109° F. −78° C. (sublimes)	Yes	830	green
Cyclopropane (C_3H_6) (Trimethylene)	2.4	10.4	928° F. 498° C.	1.45	−29° F. −34° C.	No	75	red
Ethane (C_2H_6)	3.0	12.5	959° F. 515° C.	1.04	−128° F. −89° C.	No	530	red
Ethylene (C_2H_4) (Ethene)	3.1	32.0	842° F. 450° C.	0.97	−155° F. −104° C.	Yes	1,200	red

TABLE 5–1. PRESSURIZED INDUSTRIAL GASES (CONT.)

	LOWER FLAMMABLE LIMIT (% BY VOL. IN AIR)	UPPER FLAMMABLE LIMIT (% BY VOL. IN AIR)	IGNITION TEMP.	VAPOR DENSITY (AIR = 1.0)	BOILING POINT	WATER SOLUBLE	TYPICAL CYLINDER PRESSURE (PSI)	COLOR OF REQUIRED DOT LABEL
Helium (He)	(Nonflammable; will not support combustion.)			0.14	−452° F. −269° C.	Very slightly	2,000	green
Hydrogen (H_2)	4.0	75.0	1085° F. 585° C.	0.07	−422° F. −252° C.	Slightly	2,000	red
Methane (CH_4) (marsh gas; cooking gas)	5.3	14.0	999° F. 537° C.	0.55	−259° F. −162° C.	No	2,000	red
Nitrogen (N_2)	(Nonflammable; rarely supports combustion.)			0.96	−320° F. −196° C.	Yes	2,200	green
Nitrous Oxide (N_2O) (laughing gas)	(Nonflammable; will support combustion.)			1.52	−129° F. −89° C.	Yes	800	green
Oxygen (O_2)	(Nonflammable; will support combustion.)			1.1	−297° F. −183° C.	Slightly	2,200	green

TABLE 5–2. IDENTIFYING COLORS OF GAS CONTAINERS

COLOR	GAS
Green	Oxygen
Light Blue	Nitrous Oxide
Brown	Helium
Orange*	Cyclopropane
Brown and Green	Helium and Oxygen
Red	Ethylene or Hydrogen
Gray	Carbon Dioxide or Carbon Dioxide and Oxygen

*In hospitals, cyclopropane is often found in chrome-plated cylinders with orange labels or tags.

The Department of Transportation (DOT) requires certain labels when cylinders of compressed gas are shipped between states. This labeling system is a bone of contention between the DOT and the NFPA, which prefers its own color coding system for hazardous materials. (See NFPA Standard #704.) Indeed, it seems a green DOT label can cover a multitude of hazards. The presence of a green DOT label is of small help in determining the hazards of an unknown, escaping gas.

TABLE 5–3. DOT GREEN LABEL

GAS	FLAMMABLE?	TOXIC?	SUPPORT COMBUSTION?
Chlorine	No	Yes	Yes
Oxygen	No	No	Yes
Sulfur dioxide	No	Yes	No
Ammonia	Yes	Yes	No
Carbon dioxide	No	Somewhat	Rarely
Helium	No	No	No

A red DOT label at least tells us that the gas is flammable, although, once again, *no differentiation is made between toxic and nontoxic gases*. However, if the material in a package has more than one hazard classification, the package must be labeled for each hazard. A poisonous gas must include a poison gas label. This is in accordance with the DOT general guidelines on the use of labels, Ref. Title 49, CFR, Sec. 173.403(a). Fluorine, hydrogen, hydrogen sulfide, acetylene, cyclopropane, ethylene, methane, and other gases are all shipped under the red DOT label. Their properties and reactions in a fire are as different from one another as those with a green label. Hydrogen cyanide and some other gases are shipped under a DOT poison label.

DOT label specifications indicate that each diamond (square-on-point) label prescribed must be at least 4 inches (101 mm) on each side with each side having a black solid line border ¼ inch (6.3 mm) from the edge.

The flammable gas label shown above has a red background with the printing and symbol in black.

FIGURE 5–3. DOT Flammable Gas Labels

Some gases have immediately identifiable odors. Ammonia is unmistakable. So are chlorine, sulfur dioxide, and the garlic reek of acetylene. Other odors can be faint, unfamiliar, or masked by the smell of smoke. The bitter almond odor of hydrogen cyanide may kill you if you try to inhale enough of it to identify it. Hydrogen sulfide, H_2S, has a rotten egg odor, but if a person is exposed to lethal concentrations of the gas, the olfactory nerves are immediately paralyzed and the gas will appear to have no odor at all. All in all, identification of gases by odor is of very limited value.

But, let us assume that we have somehow found out what gas confronts us. The answers to the next questions immediately become easier.

The new flammable gas placard, required when a shipment has a gross weight of 1000 pounds (454 kg), must be red with white symbol, inscription, and ½-inch (12.7-mm) border.

FIGURE 5–3. DOT Flammable Gas Labels (cont.)

Hazardous Properties

A gas can support combustion, be flammable, unstable, explosive, corrosive, toxic, or combine some or all of these unpleasant traits. Although acetylene and natural gas (methane) are always in the fire-cause top ten, gases generally cause fewer fires than flammable liquids. This is only because they are less common. If anything, flammable gases as a group are more dangerous than liquids, not only because of their properties, but because fire fighters are not as familiar with the way they will act.

Flammable gases have no flash point. They are ready to burn without preheating of any kind. Their flammable limits are generally wider than equivalent liquids. Table 5–4 compares the flammable ranges of five hydrocarbon gases, liquefied gases, and liquids.

TABLE 5–4. FLAMMABLE RANGES OF TYPICAL HYDROCARBON GASES AND LIQUIDS

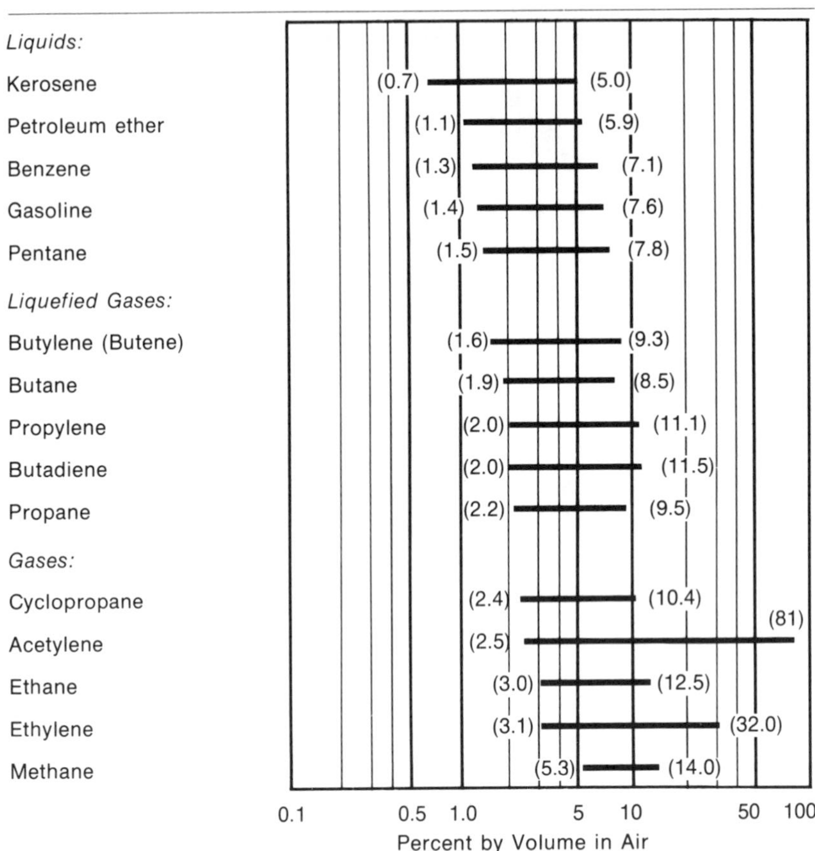

Notice that the liquids have lower explosive limits than the gases, but become "too rich" much faster. Only a few liquids can compare with the flammable range of acetylene. Gases differ from flammable liquid vapors in another important respect: All vapors are heavier than air. Some gases are lighter; some heavier. This means that explosive concentrations of certain gases (hydrogen is the perfect example) can collect in the elevated portions of confined areas. Restriction of ignition sources and venting requirements are entirely different for gases and liquids. Table 5–1 gives the fire characteristics of six flammable gases.

Pressurized Gas Cylinders

In the introduction to this chapter, we explored the connection between temperature and pressure and how quickly a fire can raise the interior pressure of a gas cylinder to dangerous levels. We talked about the way gas pressures are increased. As fire fighters, we are often called upon to reverse this process, to reduce pressures to safe levels. This can be accomplished in three ways.

1. Increase the volume of the container. Since the only way a steel container can increase its volume is by bursting, this is the least desirable way to reduce pressure.

2. Lower the temperature and pressure of an exposed cylinder by applying a cooling water spray or by physically removing the container from a threatened area.

3. Reduce the amount of gas within a cylinder by pressure relief devices.

Most cylinder valve assemblies have frangible (breakable) discs, designed to break at pressures below the bursting point of the cylinder. While the escape of flammable or toxic gases in a fire situation is a gloomy prospect, it is the better of two evils. Poisonous gases may or may not have these relief devices. Chlorine and sulfur dioxide cylinders generally have them, ammonia sometimes does, while hydrogen cyanide does not. Acetylene cylinders, as we shall see, have a unique system of pressure relief.

Another hazard associated with high-pressure cylinders is "rocketing." This occurs after accidental rupture or when a valve assembly breaks off. Driven by its contents, the cylinder becomes a missile. Cylinders take a little time to build up speed, but once launched, have been known to go through automobiles and concrete walls. The recommendation in emergencies of this type is, GET OUT OF THE WAY. Table 5–1 also gives typical cylinder pressures.

Controlling Gas Emergencies

Fortified with knowledge of the type of gas, its hazards, and the dangers associated with the cylinders which contain it, you are ready to deal intelligently with your emergency.

First of all, protect people. If a toxic or unignited flammable gas is escaping, station fire fighters to the windward, if possible, and evacuate everyone downwind. Depending on atmospheric conditions, a single cylinder can make a wide area unhealthy, especially if the gas is heavier than air. This is a good time to remember that water-soluble gases can be diluted and many other gases pushed around with fog patterns.

It will also be helpful to remember which gases are lighter than air (vapor density of less than 1.0). Memorization of the mnemonic, "HA, HA, MICE!," will help.

H, helium

A, acetylene

H, hydrogen

A, ammonia

M, methane

I, illuminating gas
(a gas mixture of ethane and methane)

C, carbon monoxide

E, ethylene

Not included in the list is nitrogen. All other gases are heavier than air.

Removing Fuel Whether the flammable material is a solid, a liquid, or a gas, one of the basic principles of the fire rectangle is the removal of fuel. When you cool a solid below its ignition temperature, you are not only removing heat, but also stopping the production of flammable vapors. This is also true when you cool a flammable liquid below its flash point. But flammable gases have no flash point and require no preheating. Shutting off their flow generally requires some physical action on your part: closing a valve, driving a plug, patching a leak. This is your major objective. If you extinguish a gas fire without stopping the flow, you convert a bad situation into one that is worse. The escaping gas will quickly form an explosive mixture with air and go in search of an ignition source. Almost certainly, it will find one.

Keeping the Gas Confined Shutting off the flow is the only sensible way to handle the situation. Sometimes it may be necessary to gain control of the flames so that a well-protected fire fighter can gain access to valves that shut off the gas supply. Make sure you know what you are doing and that you have the proper valves. It would be preferable by far to allow the fire to continue and burn out rather than increase the flow or start another by turning the wrong valves. Keep the situation cool by using fog streams to absorb heat and lessen the danger to nearby cylinders, piping, and assorted combustibles.

If extinguishment *must* be accomplished, CO_2, dry chemical, or water spray can all be effective in particular cases with particular gases. A pressurized gas can blow through a foam blanket or dislodge your attempt at a covering CO_2 cloud. As we shall see, there are some gas fires which cannot be extinguished by any known method.

Natural Gas (Methane)

When last we visited the paraffin series, we concentrated on the larger molecules from pentane upwards, looking at the hazards of gasoline, kerosene, naphtha, and the other common petroleum fractions. The first four members of this series, however, are flammable gases. We will discuss METHANE and ETHANE in this chapter, PROPANE and BUTANE in the next.

General Properties of Methane

Methane is potentially explosive and causes many fires. But its toxic hazards may be overestimated almost as much as the explosive hazards of gasoline are underestimated. Murder accomplished by turning on a modern gas fixture is pure fiction. Methane, although not toxic, can kill by suffocation because it displaces atmospheric oxygen. To be a threat to life, methane must reach a 15 percent concentration in air. Any really toxic gas, such as carbon monoxide, formed in large quantities when methane burns, is many times more deadly. It would be more efficient if a fictional murderer allowed the gas fixture to burn and closed the vent. A plastic bag over the victim's head would exclude oxygen much better than methane from an oven.

Perhaps methane gets part of its bad reputation from its various alternate names. The dreaded "fire damp" of miners is caused by pockets of methane trapped in the coal beds until set free by mining operations. In addition, methane is often produced by decaying vegetation, bubbling up in slimy swamps and such, where it is called "marsh gas." Scientists estimate that the intestinal venting of animals adds 45 million tons of methane to the air each year.

If methane were used by industry only in its pure form, it would be of relatively minor importance, just another flammable gas with various applications. Methane, however, is the principal component of natural gas. In 1967, the total production of natural gas in the United States was over seventeen trillion cubic feet. Part of this enormous volume of gas is stored under pressure or liquefied (see Chapter 6); the biggest part, however, is used to supply the daily demand of millions of homeowners and industrial users. The hazards of methane have become a matter of daily importance to every fireman.

In 1963, the Pacific Coast Gas Association published a pamphlet, entitled "Emergency Control of Natural Gas," directed to the fire service. This section on methane will draw heavily from this interesting and authoritative source. Additional comments, when they appear necessary, are in brackets [ ].

Natural gas is only 65 percent as heavy as air. This means it will rise and diffuse rapidly when it escapes in an open area. When confined in a closed room, the gas will rise to ceiling level. The air in the room will be displaced from the top *downward.* Remember this when you ventilate a room: open the windows from the top.

[Whether pure methane or natural gas is piped into your home, the hazards are similar. The limits of flammability for natural gas are four percent and 14 percent. The ignition point is quite high, about 1100 to 1200°F. (593.3 to 648.9°C.). This temperature is reached by pilot lights, flint sparks, matches, or sparks from electrical switches. One homeowner, arriving home after a week's vacation, opened his front door, switched on the lights, and was blown from the porch onto the front lawn when his entire house exploded. The house had filled with gas as the result of a slow leak, and the light switch spark served as the point of ignition. For this reason, some states require mercury switches in all buildings of public assembly.]

Notice that these figures for vapor density, flammable limits, and ignition temperature differ slightly from those for methane. Natural gas also contains trace quantities of ethane, propane, butane, and carbon dioxide. Amounts vary with the locations of the gas field. Their presence changes the properties of natural gas.

Burning natural gas produces little smoke, but it does produce a very high radiant heat. Combustibles must be wetted down with a spray to prevent their ignition by this radiant heat. [Natural gas generally burns blue with an occasional orange flash. A yellow flame may mean that soot and excess carbon monoxide are being formed because of faulty combustion. In a wall furnace, this is almost always caused by lint and dust, clogging the burners.]

Natural gas is odorless in its natural state; therefore, gas leaks in the area where it is produced or at pumping or storage stations may have no odor. Usually, the odorant that lets us detect the presence of one percent in a given volume of air is added at the distribution point. One percent is far enough below the lower flammable limit of four percent to provide a margin of safety in case of a leak. [A good odorant is nontoxic, noncorrosive, not soluble in water, and chemically inert.]

Emergency Encounters with Gas

Gas Escaping Outside If gas is escaping from the ground, an excavation, an open pipe, a manhole, a sewer, or a vault, clear a safe area around the location and barricade or rope it off. Extinguish all open flames, prohibit smoking, and make certain that electrical switches or similar possible ignition sources are not operated. Check surrounding buildings, including basements in particular, for any presence of gas odors. It may be necessary to restrict or reroute all traffic until the gas workers can bring the gas flow under control.

When gas is escaping from a broken pipe that is accessible, a wooden plug can be used to stop the leak as a temporary measure. Plugs made from redwood and pine are used because they are soft and will conform to an irregular shape caused by pipe that is not cleanly broken. They should be driven into place with a rubber mallet. If a fire fighter uses a wooden plug to make an emergency stopoff, he should tell the gas worker so that a safe, permanent repair can be made.

The gas worker can help you by supplying specific information, identifying the escaping gas and its combustibility and tracing its source. If you have to enter a manhole or vault, check to see if a concentration of combustible gas is present. If not, enter only if standby assistance is available, and keep in mind that you may be entering an oxygen deficient atmosphere. You can usually vent manholes or vaults by temporarily removing their covers. [If possible, gas company personnel and the fire fighters should stand upwind of a leak.]

Gas Burning Outside If gas is burning outside, the fire fighter should make no attempt to extinguish the fire. Burning gas will not explode. Clear the danger area and barricade or rope it off. *Never operate gas valves in the street in this type of emergency.* Turning the wrong valve can create an emergency in another area worse than the one at hand and further endanger life or property. Spray water mist on any surrounding combustibles if they are in danger of igniting. Do not use water on burning gas at its point of escape. If this point is in an excavation, the hole becomes filled with mud and water which makes the repair slower and more hazardous.

Whenever or wherever gas is involved, immediately call the gas company. Its personnel are instructed to report their presence to the fire officer in charge upon arrival.

Gas Escaping Inside If gas is escaping inside a building, ventilate the area, starting where the gas concentration is strongest. If gas is escaping in quantity, clear the building of its occupants. Shut off open flame devices by operating manual controls, but *do not* operate electrical switches. [This includes the main electrical switch. Opening this switch to cut off electrical sources of ignition causes a tiny spark at each switch inside the building. This has been the cause of many explosions.]

In some cases, the fire officer in charge may determine that it is necessary to shut off the gas to the building at the service valve. If you turn off a valve, leave it off and tell the gas worker, who should decide the proper time to turn it on again. [Locating a gas leak either indoors or outdoors can be troublesome. Sometimes, leakage of gas into the ground causes visible damage to vegetation. The leak can be located by noticing dry, discolored plants.]

Gas Burning Inside If gas is burning inside a building, shut it off at the meter or outside at the curb valve. If the gas supply cannot be safely shut off, keep the surrounding combustibles wet with spray streams until the gas company emergency crews can control the flowing gas. If it appears that inside piping or installations are threatened by a fire inside the building, whatever the origin, the fire officer in charge should determine if it is necessary to turn off the gas.

[As a matter of practice, many fire departments turn off the gas in a burning building until it can be determined that no hazard of gas leakage exists.]

A Case History

While emergencies involving natural gas, principally methane,* are not rare, the wonder is not how often it is involved, but how seldom. Seventeen trillion cubic feet is an incomprehensible amount. Sometimes things are bound to go wrong. The following is a summary of a report by the AIA entitled, "The Brighton Gas Fire and Explosion Catastrophe."

> On the afternoon of September 21, 1951, in Brighton, New York, an explosion, apparently resulting from an accumulation of gas, took place in a regulator vault of the gas distribution system. As a result, regulating valves were pushed into the wide-open position. Gas, at 30 pounds pressure, flowed into a low-pressure system designed for pressures of less than one pound. The effect over a 25-block area was sudden and disastrous.
>
> Where burners or pilot lights were in operation, torch-like flames roared up to a height of two feet or more. Some stove burners lit, although the gas was shut off. In other cases, the pilots were blown out and unburned gas escaped. Flame came out of automatic water heaters and completely enveloped the tanks. Furnaces flared into life. More than 12 percent of the meter cases failed in an area containing 1500 homes. A number of failures caused the formation of explosive mixtures within the buildings.
>
> A few minutes after the failure of the regulators, the first home exploded, killing two children and severely burning a woman. At short intervals thereafter, other houses throughout the area were shattered by explosions which, in most cases, completely demolished the buildings.
>
> Fire broke out in the splintered debris of many of the buildings and extended to adjacent dwellings. Still other fires resulted from the

*Ethane, by comparison, is almost nonexistent. It is shipped in pure form inside steel cylinders for use as a refrigerant, a fuel, or in the production of organic compounds. Ethane is colorless and odorless, a nice, uncomplicated flammable gas which will explode if you give it half a chance. Table 5–1 gives its fire characteristics.

abnormal operation of gas appliances. Within 90 minutes, 33 homes were totally destroyed or seriously damaged, 14 more were damaged to some degree.

Brighton had three engine companies. Two were involved in trying to control the fire at the regulator vault. Until it received assistance from 36 mutual aid companies from surrounding cities, it was up to the third Brighton company to respond to 47 explosions and fires.

A central pool of fire equipment was set up. In some cases apparatus was sent out as explosions were heard, rather than waiting for alarms to be received through more normal channels. The worst was over about three hours after the original explosion. The explosions had stopped and all fires were under control. The gas company had stopped the flow of gas. Not surprisingly, there were a few rekindles.

Fortunately, most people had left their homes. They did not return until hours after the danger was over. For this reason, no additional fatalities or serious injuries were suffered as a direct result of the explosions, although one woman died of a heart attack brought on by the excitement.

The life loss was also kept to a minimum by the occurrence of this disaster during an afternoon period when most of the children were at school, which was evacuated, and a great many people were away from home shopping or working. If this had occurred during the night, the life loss would have been far greater.

The activities of the fire department were also of great importance. They spread warnings and went from house to house to make certain the gas was shut off. One fireman alone turned off gas in more than forty houses.

Anesthetics

When doctors use an anesthetic, a delicate balance must be maintained between the depth of unconsciousness necessary and the medical effect the anesthetic will have upon the patient. All our vital organs—heart, brain, lungs—are regulated by nerve impulses. There must be enough anesthetic to deaden pain but not so much as to interfere with life functions. Diethyl ether was a popular anesthetic in the past, but many preferred anesthetics have been developed over the last ten years.

Divinyl ether is more powerful than ethyl ether, but it causes occasional ill effects and has the troublesome habit of causing ice formation in masks.

ETHYLENE, one anesthetic in use today, is a colorless, flammable gas with a pleasant, sweet odor. It has an important role in the manufacture of organic compounds, ripens fruit and gives them color, and increases the growth rate of seedlings. We will meet ethylene again in a later chapter when we consider how plastics are made. Figure 5–5 diagrams its structure; Table 5–1 lists its fire characteristics.

Remember: Ethylene will cause unconsciousness on the fire scene as well as in the operating room. *Note:* Ethylene, in sunlight, is spontaneously explosive with chlorine gas. Storage of these two gases should be separated.

```
    H   H
     \ /
      C
     / \
 H—C — C—H
   /     \
  H       H
```

FIGURE 5–4. Cyclopropane (C_3H_6)

CYCLOPROPANE is another colorless gas. It is a strong anesthetic. Concentrations exceeding one part in 2500 in air (400 parts per million) can cause unconsciousness. Cyclopropane is shipped in steel cylinders under fairly low pressure. It is another example of the varied structures that the ever-versatile carbon atom will build: this time, triangles.

ETHYL CHLORIDE is not a hydrocarbon. Each molecule contains a single chlorine atom. Occasionally, ethyl chloride is employed as a general anesthetic by doctors, dentists, and veterinarians. More often, it is used for local anesthesia and is applied as a spray. Since ethyl chloride boils at 54°F. (12°C.), it turns rapidly to vapor, absorbing enough heat from the human tissue it touches to freeze and deaden the area.

Ethyl chloride is also a refrigerant and a solvent, and is used in the manufacture of tetraethyl lead. Concentrations over 4 percent (40,000 parts per million) can be fatally anesthetic. It burns with a green flame, forming highly toxic phosgene gas. The flash point is extremely low: −58°F. (−50°C.).

Despite its sweet odor, prolonged inhalation of NITROUS OXIDE can be fatal. It is used only in operations of short duration. Even then, "laughing gas" can have a peculiar influence on patients. While nitrous oxide is not flammable, it contains a higher percentage of oxygen than air and supports combustion more readily. When flammable anesthetics are mixed with nitrous oxide or oxygen their explosive potentiality is increased.

Storage and Handling of Anesthetics

Most anesthetics in use today are flammable gases or volatile liquids used in a highly dangerous form, but their use is unavoidable at the present time. If trouble occurs, there is an immediate hazard to life.

The only possible solution lies in a combination of stringent regulation of storage and handling, the complete elimination of ignition sources, rigid observation of safety rules, and our proficiency in inspection and fire fighting.

TABLE 5–5. RELATIVE FLAMMABILITY OF ANESTHETIC MIXTURES

	FLAMMABLE LIMITS IN AIR (PERCENT BY VOLUME)		FLAMMABLE LIMITS IN NITROUS OXIDE (PERCENT BY VOLUME)		FLAMMABLE LIMITS IN OXYGEN (PERCENT BY VOLUME)	
	LOWER	UPPER	LOWER	UPPER	LOWER	UPPER
Cyclopropane	2.4	10.4	1.6	30.3	1.8	60.0
Ethyl chloride	3.8	15.4	2.1	32.8	4.0	67.2
Ethylene	3.1	32.0	1.9	40.2	2.9	79.9
Ethyl ether	1.9	48.0	1.5	24.2	2.0	82.0
Vinyl ether	1.7	27.0	1.4	24.8	1.8	85.5

Flammable anesthetics should be stored together in a cool, well-ventilated place outside the area of use. The location should be clearly designated. The following provisions of the National Electrical Code should be enforced.

1. Any room or space in which flammable anesthetics or volatile flammable disinfecting agents are stored shall be considered a Class 1, Division 1 location throughout.

2. In an anesthetizing location, the entire area shall be considered to be a Class 1, Division 1 location which shall extend upward to a level five feet above the floor.

Briefly, the Code separates hazardous occupancies into three classes.

Class 1 includes flammable gases or vapors.

Class 2 includes combustible or explosive dusts.

Class 3 includes easily ignitable fibers.

Class 1 is further separated into four groups.

Group A: atmospheres containing acetylene.

Group B: atmospheres containing hydrogen, or gas and vapors of equivalent hazards.

Group C: atmospheres containing ethyl ether vapor, ethylene, or cyclopropane.

Group D: atmospheres containing most of the common flammable vapors or gases.

Each of the classes has two divisions. Division 1, in each case, has a higher degree of hazard because there is a greater possibility that a hazardous atmosphere will form for one reason or another. Therefore, the requirements for approved electrical equipment are stricter.

Class 1, Division 1 locations must have explosion-proof electrical equipment without exception. The five-foot requirement takes into consideration the vapor density of ethyl ether.

Even with explosion-proof equipment, static electricity remains a possible source of ignition. Although tightly controlled by grounding, humidifying the air, and special clothing and shoes worn by doctors and nurses, the generation of static is still the primary cause of operating room explosions. A complete survey of the problem is contained in NFPA Pamphlet 56, "Standards for the Use of Flammable Anesthetics."

Fires involving ethylene, cyclopropane, and ethyl chloride above 50°F. (10°C.) are best fought like those involving all other flammable gases: shut off the flow. None of these gases is highly soluble in water. It would be difficult to wash them from the air. Use spray streams to keep the cylinders cool. If a burning cylinder is mounted on an anesthetic machine, move the cylinder to a safe place. Do not overlook the oxygen bottle which may be alongside. If its pressure relief device fails, burning rate will be increased suddenly. When possible, try to separate the cylinders. If extinguishment is required, dry chemical or CO_2, supported by spray streams, should be effective.

Remember: These are anesthetics. Wear self-contained gas masks.

Finally, as one vetran anesthetist put it, "The fire after an operating room explosion is generally not too severe. The damage has already been done."

Acetylene

If ever the properties and uses of a compound were dictated by its structure, that compound is acetylene, which has within it two carbon atoms joined by a triple bond. Let us see what this looks like and what it means to a fire fighter.

Figure 5–5 shows the structural formulas of three hydrocarbon gases that contain two carbon atoms. One is ethane, a member of the paraffin series. Its carbon atoms are linked by a single bond. (All the other bonds are saturated by hydrogen atoms.) Ethylene doubles this linkage and acetylene triples it.*

---

*Both ethylene and acetylene are the first members of two series of unsaturated hydrocarbons: the olefins and the acetylenes. The olefins take the suffix *-ene* and include propylene (propene) and butylene (butene). The acetylene series is not well known. Members have the suffix *-yne*. They include propyne and butyne. Acetylene should be called ethyne, but it is not, has not been, nor will it be.

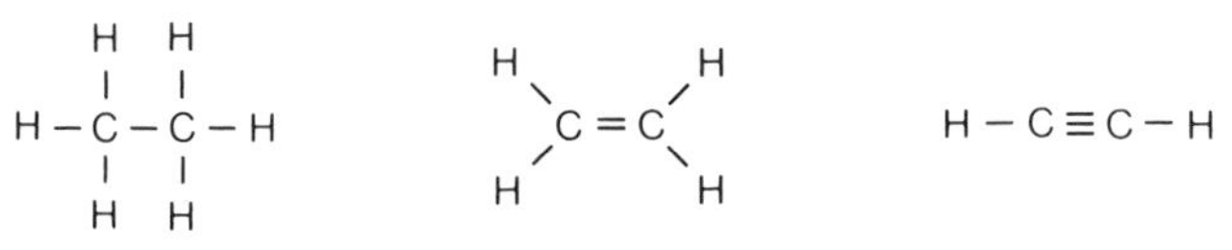

H H
| |
H−C−C−H
| |
H H

Ethane
C_2H_6

H H
\ /
C=C
/ \
H H

Ethylene
C_2H_4

H−C≡C−H

Acetylene
C_2H_2

FIGURE 5–5. Hydrocarbons with Two Carbon Atoms

The double bond of ethylene contains a great deal of energy. Under the right combination of pressure and heat it can be broken and ethylene reformed into new products. A triple bond, such as the one in acetylene, is more than just reactive and energetic. It is unstable and potentially explosive. But acetylene has many unique properties which industry can put to work.

Use and Production of Acetylene

When we think of acetylene our first thought is of oxy-acetylene welding and cutting. Some large fires have been caused by the bouncing sparks from a cutting torch. Many of these losses could have been avoided by moving the job away from combustibles, by moving the combustibles, or by shielding one from the other. Yet, three-quarters of the acetylene produced in this country is used for purposes other than welding or cutting.

Acetylene can be found somewhere in the ancestry of a wide variety of materials; synthetic chemicals, such as acetaldehyde, acetic acid, and acetone; water-based paints; dry-cleaning solvents; and such plastics as Neoprene, Orlon, polyvinyl chloride, Lucite, and Plexiglas. When supplied with the correct amount of air, acetylene burns with a white light instead of its usual smoky flame. This light has many similarities to pure sunlight and can be used in place of electricity, such as in navigation buoys and the carbide lamps of miners. With all these uses, and many more not mentioned, it has been estimated that within the next decade the yearly consumption of acetylene will approach twenty billion cubic feet.

Acetylene is produced in two ways; by the action of water on calcium carbide, and by the cracking and reformation of other hydrocarbons. Both processes are economically feasible; choice depends on where the plant is located. Where hydroelectric power for electric furnaces is cheap, the carbide process is favored. In oilier portions of the country, acetylene is produced directly from natural gas and propane by "thermal cracking."

When coal and unslacked lime are heated in electric furnaces, calcium carbide and carbon monoxide are formed.

$$\underset{\textbf{Coal}}{3C} + \underset{\textbf{Unslaked lime}}{CaO} \xrightarrow{\text{heat}} \underset{\textbf{Calcium carbide}}{CaC_2} + \underset{\textbf{Carbon monoxide}}{CO}$$

Acetylene is then generated by the action of water on the carbide.

$$\underset{\textbf{Calcium carbide}}{CaC_2} + \underset{\textbf{Water}}{2H_2O} \longrightarrow \underset{\textbf{Slaked lime}}{Ca(OH)_2} + \underset{\textbf{Acetylene}}{C_2H_2}\uparrow$$

Inside large metal generators, water and carbide are brought together. Two processes are used: water-to-carbide and carbide-to-water. The latter is preferred because the excess water helps to dissipate the heat caused by the reaction. Acetylene gas is formed, drawn off, purified, and compressed into cylinders. Occasionally, acetylene generators explode from the accidental overpressures, from the ignition of an acetylene-air mixture by glowing carbide or other sources, or from sudden violent decomposition. (See discussion of calcium carbide in *Explosive and Toxic Hazardous Materials.)*

Because of the great saving in shipping cost, most commercial manufacturers ship acetylene in the form of the carbide, rather than as gas in cylinders. Each 100 pounds of carbide will generate 31 pounds of acetylene, a pound yielding 4.6 cubic feet of gas. Compare this with five pounds of acetylene in a cylinder with a shipping weight of 100 pounds.

Acetylene Hazards

Pure acetylene is an odorless, colorless gas. Commercial acetylene contains impurities—phosphine, ammonia and hydrogen sulfide—which give it a garlic odor. Overall, the toxic hazard of acetylene is relatively slight. While it can have an anesthetic effect in high concentrations, it functions mainly as an asphyxiant.

But there is nothing slight about acetylene's flammable hazard. It has a wider explosive range than any other common flammable gas: 2.5 percent to 81 percent. With oxygen, the upper flammable limit rises to 93 percent. Ignition temperature is remarkably low for a gaseous hydrocarbon, 571°F. (299°C.). And acetylene burns hot! An oxy-acetylene flame has a temperature of about 5700°F. (3149°C.), the highest temperature of any known mixture of combustible gases, a temperature far higher than the flame of an ordinary flammable liquid or solid. (See Table 5–6).

TABLE 5–6. APPROXIMATE FLAME TEMPERATURES OF CERTAIN FLAMMABLE GASES

Methane (in air)	3407°F. (1875°C.)
Butane (in air)	3443°F. (1895°C.)
Ethane (in air)	3443°F. (1895°C.)
Propane (in air)	3497°F. (1925°C.)
Hydrogen (in air)	3718°F. (2048°C.)
Carbon monoxide (in air)	3812°F. (2100°C.)
Acetylene (in air)	4217°F. (2325°C.)
Hydrogen (in oxygen)	4820°F. (2660°C.)
Carbon monoxide (in oxygen)	5397°F. (2981°C.)
Acetylene (in oxygen)	5710°F. (3154°C.)

Acetylene is also unstable. If severely shocked or ignited at pressures above 15 psig, or subjected to fire temperatures, the triple bond can break apart (decompose), forming carbon and hydrogen. Once started, decomposition does not require the presence of air to continue and can quickly escalate into explosive violence. As a consequence, the NFPA forbids the generation and use of free acetylene at pressures higher than 15 psig (a little more than two atmospheres). How then can acetylene cylinders have a standard storage pressure of 250 pounds?

Those who have seen the interior of an acetylene cylinder know the answer: The inside is a calcium-silicate filler made from sand, lime, and asbestos. This filler looks absolutely solid, but is 92 percent nothing. It prevents the formation of pockets of free acetylene. In addition, each cylinder is filled with acetone before acetylene is pumped into it. Dissolved in acetone, acetylene is stable. For each atmosphere of pressure, acetone absorbs 25 times its own volume of acetylene. At the standard charging pressure of 250 pounds (about 16 atmospheres), a cylinder contains 425 times its volume of acetylene. In summary, a modern acetylene cylinder contains an almost solid-looking filler, whose microscopic pores are filled with acetylene dissolved in acetone. Were it not for this, acetylene would not be an important industrial gas. It would be too dangerous to use.

Acetylene cylinders have another safeguard. Instead of frangible discs, acetylene cylinders have soft plugs designed to melt around the boiling point of water. Depending on the size of the cylinder, there may be one to four of these soft plugs. All of these specialized safeguards and hazardous properties can cause a variety of fire problems when acetylene is involved.

1. When one of the soft plugs melts, a sudden jet of white flame may extend a dozen feet into the air as acetylene and acetone vapor escape.

(This flame will diminish in less than half an hour even though the bottle is full.) If a large number of cylinders are stored together, it can look like Cape Canaveral as all the soft plugs melt. This is the time for discretion, the use of deluge sets for the protection of exposures, and the full support of sprinkler systems.

2. When in use, acetylene is almost always stored next to oxygen. The two cylinders may even be chained to each other. If the acetylene bottle is on fire and the pressure relief device on the oxygen bottle is still intact, direct every effort toward keeping the setup cool with the use of spray streams. Otherwise, there is the possibility of explosive rupture of the oxygen cylinder or the sudden appearance of a very large oxy-acetylene flame when its frangible disc bursts.

3. Decomposing acetylene can propagate back into its cylinder. The cylinder may overheat to the point where the metal fails. Unless a cooling spray is kept upon it, it will probably explode.

4. When bottles are knocked over, there is also the possibility of a running acetone spill fire if the soft plug melts. This may complicate matters just enough to make the situation impossible. If this happens to a single bottle, the use of dry chemical, supported by fog, can quickly extinguish the acetone fire.

Basic fire-fighting techniques for a flammable gas still hold true for acetylene, although the problem of how to shut off the flow may get a little complex. *An acetylene fire should not be extinguished except to facilitate the immediate shutting off of the flow of gas.* Be careful about moving a heated cylinder, even if the fuse plug has not melted. If a leak is not on fire, it may be possible to stop the flow in some manner and get the cylinder outside.

The storage area of acetylene must have Class 1, Group A electrical equipment. For obvious reasons, cylinders must be stored upright in a cool, well-ventilated area, preferably protected by a deluge sprinkler system.

Finally, just to prove it is a triple threat, acetylene also forms explosive compounds with copper, silver, mercury, and chlorine. It should never be stored near chlorine, and piping containing acetylene should always be made of iron or steel. The only time acetylene should touch copper is in the head of a torch.

Hydrogen

There are eleven elemental gases. Seven are considered more or less inert: helium, neon, argon, krypton, xenon, radon, and nitrogen. Three will support combustion: fluorine, chlorine, and oxygen. Only one, hydrogen, is flammable. Indeed, its flammable range of 4 percent to 75 percent is almost as wide as that of acetylene. Hydrogen burns in air with a high flame temperature, 3700°F. (2037°C.); it burns even hotter when part of an oxy-hydrogen mixture. (See Table 5–6.) Most of the energy of burning hydrogen is released as heat rather than light. It burns with a short, intense, pale blue flame that is almost colorless and difficult to see in daylight.

Hydrogen is the lightest element. Although astronomers estimate that hydrogen makes up 90 percent of the universe, free hydrogen gas is relatively rare in the earth's atmosphere. Being only 7 percent as heavy as air, hydrogen ''leaks'' into space, for the earth does not have enough gravity to hold it. In nature, the hydrogen that remains is locked into such compounds as water and alcohols and hydrocarbons; hydrogen gas is largely a man-made product.

Storing hydrogen is difficult. The hydrogen molecule, H_2, formed by two hydrogen atoms, is smaller and lighter than any other single atom. As a consequence, hydrogen likes to leak; it will find the tiniest opening in a pipe or container. Pipe threads and valve stem packings must be tight. A high-pressure hydrogen leak, such as from a cylinder at 2000 psi, can ignite spontaneously. Ignition may be caused by the friction of escaping molecules.

All flammable gas leaks are dangerous but especially so if, like hydrogen, the gas is odorless and colorless. In an enclosed space, hydrogen diffuses quickly and rises to the top. Although explosive mixtures are rapidly formed, if the leak is small, hydrogen may dissipate so rapidly that an explosive concentration may not be reached. But, since it burns with an almost invisible and intensely hot flame, and has a wide explosive range that makes for extreme ease of ignition, the hazards of hydrogen gas should never be underestimated. Hydrogen is not toxic and, since asphyxiating concentrations should only be found near the ceiling, its health hazard is not considered severe.

Hydrogen is produced industrially in several ways: by passing steam over hot coal or coke to form WATER GAS, a mixture of hydrogen and carbon monoxide; by thermal decomposition of natural gas; by breaking apart the ammonia molecule; or by passage of an electric current through water (electrolysis). The increasing use of liquid hydrogen will be discussed in the next chapter.

Fire fighters should know that certain acids, when reacting with metals, or when heated, release hydrogen. This can be troublesome if pressure builds up in drums. Hydrogen is also released by the negative elements of lead storage batteries while they are being charged. An investigation by the U.S. Bureau of Mines showed that flammable concentrations of hydrogen existed in a percentage of the battery rooms tested. Many small fires and explosions have occurred when the gas ignited. In some cases, batteries have blown apart, scattering the acid. Adequate ventilation of battery rooms near the ceiling level is important.

Hydrogen is used in welding as part of the oxy-hydrogen flame;* for the hydrogenation of vegetable oils (to turn Wesson Oil into Spry, add a cup of hydrogen); for the synthetic production of ammonia and methanol; and for the cooling of large electrical generators. (Since the Hindenburg disaster, helium has replaced hydrogen in all large manned balloons.)

Fighting a hydrogen fire can proceed as with other flammable gases. Once again, *shut it off.*

QUESTIONS ON CHAPTER 5

1. Which of the common gases has the widest flammable range?
2. What are the names of four liquids or gases used as anesthetics?
3. What is the primary objective in flammable gas fires?
4. Natural gas consists principally of what hydrocarbons?
5. What gas has an odor of garlic?
6. What are the three ways by which the pressure of a gas is increased?
7. What is the principal cause of operating room explosions?
8. What material reacts with water to form acetylene?
9. What flammable liquid is included inside pressurized acetylene cylinders?
10. What is the lightest gas known to science?
11. Beginning with a gas at normal temperature, how much of an increase in temperature will double its pressure according to the rule of thumb?
12. According to the National Bureau of Standards, an orange cylinder should contain what gas?
13. Name five gases which are shipped under a DOT green label.
14. Name three ways to shut off the flow of gas.
15. What are the toxic hazards of methane?
16. Why is carbon monoxide often formed by wall furnaces?
17. With what gas is ethylene spontaneously explosive?
18. What highly toxic gas does burning ethyl chloride form?
19. The National Electrical Code separates hazardous occupancies into three classes. What does each of these classes cover?
20. Is ethylene water soluble?

---

*Flammable mixtures of hydrogen and nitrogen are also sometimes encountered.

BIBLIOGRAPHY

American Insurance Association Bulletins:

No. 66, "Welding and Cutting as a Fire Cause" (revised June, 1956).

No. 85, "Hydrogen" (April 11, 1955).

No. 146, "Utility Gas Distributed by Utility Companies" (January 1957).

American Insurance Association Report:

"The Brighton Gas Fire and Explosion Catastrophe" (1951).

National Fire Protection Association:

Standard No. 56, "The Use of Flammable Anesthetics" (1962).

Union Carbide Corp. (Linde Division):

"The Acetylene Cylinder" (1959).

"Precautions and Safe Practices in Welding and Cutting" (1960).

Pacific Coast Gas Association:

"Emergency Control of Natural Gas" (1963).

Compressed Gas Association:

"Safe Handling of Compressed Gases" (1967).

DEMONSTRATIONS

1. If available, a cutaway model of an acetylene cylinder is most instructive. It doesn't seem possible that all that acetone and acetylene can find room inside.
2. The differences in vapor density between gases can be demonstrated by two small, compressed cylinders of helium and carbon dioxide and two toy balloons. Fill the balloons. Helium goes up; CO_2 settles to the floor.
3. Generation of acetylene:
   Equipment needed: a deep pyrex or metal dish of fairly small diameter, two candles, water, and some calcium carbide.

   In the center of the dish, place a heavy holder containing a candle. Fill the dish with water to within a few inches of the candle flame. Drop one or two pieces of carbide into the water. It will fizz, producing acetylene. The gas is immediately ignited by the candle. If the dish is too large, some unignited acetylene may escape. The blackness of the smoke and the heating of the water can be commented upon. The second candle is used to re-light the original candle that is often extinguished by sudden puffs of acetylene. Practice this first.

6 Liquefied Gases

As we know, when water turns to steam it increases in volume some 1700 times. While ratios vary to some extent, there is a similar volume relationship between other liquids and gases. A small amount of liquid will produce a great deal of gas. The industrial importance of this fact is evident. By liquefying a gas, we can store a large quantity in a small area. This is both convenient and economical.

Liquefying

One method of liquefying a gas is simply to cool it below its boiling point. Cool steam below 212°F. (100°C.), or butane below 31°F. (−0.6°C.), or ammonia below −28°F. (−33°C), or propane below −44°F. (−42°C.), and these gases turn into liquids. The liquid forms because the speed of molecular movement has slowed enough to allow the molecules to adhere.

Critical Temperature

Since it is also possible to force gas molecules closer together by applying pressure, will they adhere if enough pressure is applied? It all depends. We have seen that many gases are stored in highly pressurized cylinders. They show no tendency to liquefy. In fact, a gas like hydrogen will not liquefy at room temperature no matter how much pressure is applied. Clearly, there is a factor preventing liquefaction. This is the CRITICAL TEMPERATURE, the temperature above which it is impossible to liquefy it by pressure alone. The pressure required to liquefy a gas at its critical temperature is called the

CRITICAL PRESSURE. For instance, the critical temperature of water vapor is 705.2°F. (374°C.). The critical pressure necessary to liquefy water vapor at this temperature is around 3200 psi (218 atmospheres). Above this temperature, no amount of pressure will liquefy water vapor. As the temperature drops below critical, so does the amount of pressure required. Finally, at 212°F. (100°C.), the only pressure needed is supplied by the atmosphere: 14.7 pounds. The reason hydrogen will not liquefy, even under cylinder pressures of 2000 psi (136 atmospheres), is that its critical temperature is extremely low, close to 400 degrees below zero F. (−240°C.). In order to be liquefied, hydrogen gas must somehow be reduced to this unearthly temperature. We will see how this is accomplished in Chapter 7.

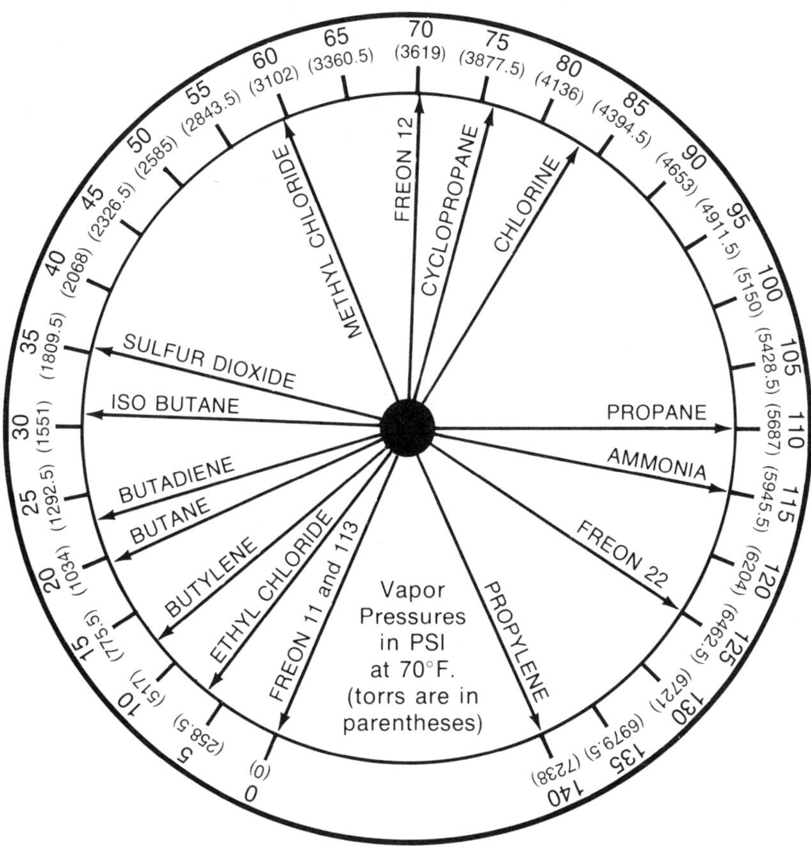

FIGURE 6–1. Vapor Pressures of Selected Gases at Normal Temperatures

Many gases, however, have critical temperatures well above normal air or room temperatures. The critical temperature of butane is 306°F. (152°C.), that of propane 206°F. (97°C.), that of ammonia 271°F. (133°C.). If put under sufficient pressure, these gases will liquefy without cooling. When put in the proper container, they are self-sustaining. They will boil and produce vapors until enough pressure builds up inside the container to prevent further production; for, as the pressure on a liquid increases, so does its boiling point.* Inside each liquefied gas container there is a liquid with a pressurized gas above it. As the gas is withdrawn for use, the pressure will temporarily lower. A little more liquid boils until the pressure is once again in line with the prevailing temperature. In this manner, a gas can be stored at temperatures above its normal boiling point for months or even years. Figure 6–1 (page 125) compares liquefied gas vapor pressure at 70°F. (21°C.).

We can readily see both the dangers and convenience inherent in liquefied gas storage. Containers must be sturdy to withstand interior pressures and external shock. If an explosive rupture occurs, all the stored pressure is released at once and the entire liquid content of the tank boils away. The amount of gas produced will be considerable. In spite of this hazard, the kinds of liquefied gas and amounts in storage are constantly increasing. The most widespread are the liquefied petroleum gases, butane and propane.

Butane and Propane: The LP Gases**

Liquefied petroleum gases have become big business. Annual production exceeds twelve billion gallons. A large percentage of this is used for a fuel gas "beyond the mains," on farms and in house trailers, as a supplement to urban natural gas supplies, and, increasingly, as a motor fuel. Tremendous quantities of butane are used in the production of high-octane gasoline. LP gases are obtained as one of the by-products of oil and natural gas wells, or by deliberate formation from other petroleum fractions.

Containers for LP gases range in size from the one-pound, hand-held cylinders often sold in department stores to large metal containers which hold thousands of gallons. A house trailer, on the move, often carries twin 10-gallon (38-liter) tanks. Another common size is the 28-gallon (106-liter)

*Critical temperatures are closely related to boiling points. Butane, with a boiling point of 31°F. (−0.6°C.) is very close to being a liquid on its own. Just a slight increase in pressure will liquefy it: 31 pounds at 70°F. (1603 torrs at 21°C.). But, the relationship between boiling point and critical temperature is subject to some variation: While ammonia and chlorine both boil at −28°F. (−33°C.), the critical temperature of chlorine is 291°F. (143°C.), twenty degrees higher than that of ammonia.

**The technical definition of a *vapor* is a gaseous material that exists *below* the critical temperature, while a *gas* exists only *above* this temperature. Properly, the LP gases should be called the LP vapors in the temperature range between their boiling point and their critical temperature. They never are, which shows how much attention is paid to technical definitions.

returnable cylinder with a liquid content of 122 pounds (55 kg). All these containers are built to withstand gas pressures generated at 100°F. to 130°F. (38°C. to 119°C.). Most containers also have pressure relief devices: spring-loaded valves or some sort of fusible plug. Spring-loaded valves are preferred because they will reseat themselves if the overpressure subsides. Once a plug goes it is gone for good. Other methods of protecting LP gas containers include burying, insulation, water-spray protection systems, gas detection and alarm systems, and staggered spacing of groups of tanks.

With propane and butane, we complete the paraffin series of hydrocarbons. From a fire fighter's viewpoint, we may have saved the worst for the last. In many ways, they are the most hazardous of the series. Although they diffuse fairly rapidly, they are heavier than air and will hang together in a cloud far longer than methane or ethane. At the same time, they are gases under some pressure. As such, they can present a more severe fire problem than liquid hydrocarbons, even such a volatile one as gasoline. The ratio of liquid to gas for propane is 1 to 270. The amount of flammable vapor propane will produce in a short time far exceeds what would be produced by

TABLE 6–1. THE PROPERTIES OF BUTANE AND PROPANE

	PROPANE (C_3H_8)	BUTANE (C_4H_{10})
Flash point (of liquid)	below −100°F. (−73°C.)	−76°F. (−60°C.)
Ignition Temperature	871°F. (466°C.)	761°F. (405°C.)
Flammable limits	2.2% to 9.5%	1.9% to 8.5%
Specific gravity of liquid at 60°F.	0.509	0.582
Vapor density	1.6	2.0
Boiling point	−44°F. (−42°C.)	31°F. (−0.6°C.)
Water solubility	No	Slight
Volume of gas per pound of liquid (at 60°F. and atmospheric pressure)	8.5 cubic feet	6.5 cubic feet
Vapor pressure at 70°F. (21°C.)	110 psig	19 psig
Vapor pressure at 100°F. (38°C.)	192 psig	59 psig
Critical temperature	206°F. (97°C.)	306°F. (152°C.)
Critical pressure	617 psig	551 psig
Weight per gallon of liquid	4.24 pounds	4.84 pounds
Flame temperature (approximate)	3497°F. (1925°C.)	3443°F. (1895°C.)

an equivalent amount of gasoline. In short, the LP gases seem to combine the worst properties of both flammable liquids and gases.

Some other properties of propane and butane require comment. They are almost odorless. For leakage detection, a strong-smelling chemical compound called a MERCAPTAN is added. Of course, we should not ignore this smell when it occurs, but we cannot depend entirely upon its presence. Some chemical reactions require the use of a pure gas, so no odorant is added. It is therefore possible to encounter an odorless LP gas. It should be so marked, but it may not be. Both gases are also colorless. However, a liquid leak vaporizes almost immediately, chilling the air, and condensing and making visible the water vapor it contains. Even though the gas is invisible, an LP gas leak can be detected by this vapor cloud. The point of leakage may also be frosted. Sometimes we can estimate the level of liquid in a leaking container by a ring of frost caused by the rapid vaporization of the liquid as it seeks to restore a pressure balance. When vaporization takes place it absorbs heat from its environment, including the liquid inside the container.

Propane systems depend upon natural vaporization within the container to maintain a steady flow of gas. The amount of heat required for vaporization depends upon the rate of use and the climate. If there is heavy usage, the container may require additional heat because the temperature of the liquid may drop close to its boiling point, but in the United States this is not usually a problem, since propane will vaporize by itself at normal temperatures throughout the country.

Propane has the three qualities needed for a successful LP gas: It is highly flammable; it can be liquefied by moderate pressure; and it reverts back to a gas at all convenient temperatures. This is not so true of butane.

Butane's boiling point, 31°F. (–0.6°C.), means it will not vaporize at many winter temperatures. Therefore, although *butane* has become a synonym for the LP gases, propane and mixtures of propane and butane are much more commonly used, along with such inevitable impurities as iso-butane, propylene, and butylene. Propane has the disadvantage of requiring heavier tanks and equipment because of its higher vapor pressure.

Probable LP Gas Emergencies

LP gas emergency situations include a leaking LP container accompanied by a vapor cloud, and an LP container (or containers) on fire, or exposed to fire. Handling of these emergencies requires these considerations.

1. Protect people. Any vapor cloud will be downwind from the leak. Firemen should approach from upwind side, if at all possible. For this reason, places using LP gas should allow access from all sides. The upwind approach is equally important if the LP containers are on fire.

Explosions from a newly created vapor cloud can occur. Remove all persons from the area of the cloud or from its probable path. Keep them back at least 2000 feet (600 meters) from the area of the cloud wherever it is or goes. The only exceptions to this rule are those people required to deal with the emergency.

Remember: Large LP tanks are *horizontal* tanks. (See Chapter 4, page 86.) Do *not* approach them from the ends.

2. Shut the gas off. This basic rule of gas fire fighting applies with even more force to handling LP gas emergencies. There is no tactic more worthwhile than shutting off the flow. Close valves, at the container or remotely, by using valve wheels or wrenches; by crushing or crimping copper tubing; or, as happened on a Hollywood freeway when a tank of butane overturned, by driving redwood plugs into holes. Consult plant personnel or drivers about the location of proper valves. If you are lucky, you may encounter the type of system where valves close automatically. If valves cannot be located or used, you will have to shut off every ignition source in the path of the vapor cloud.

Remember: An LP gas vapor cloud is *heavier than air* and will sink into low places.

3. Use water to direct the vapor. Although dry chemical will extinguish a very small LP gas fire, *there is no known method or material that will extinguish a large one.* However, the proper use of water will be of assistance. Water is absolutely indispensable. Large amounts should be immediately available. Water can help to protect fire fighters closing valves. Without fog patterns, it may be impossible to approach necessary shutoffs. Water can help disperse LP gas vapor. It will not dilute the vapor, but it can push it to a safer location. Fog patterns should be used immediately. Direct the spray across the normal vapor path. If the cloud ignites, there will be a tremendous release of radiant heat that a sufficient amount of fog can help lessen.

Several facts about vapor clouds should be kept in mind: Flames will progress at 15 feet (5 meters) per second through a large cloud, a rate which is about one-half the speed of a desperate man; running 100 yards (90 meters) in 10 seconds will allow a man to travel 30 feet (10 meters) per second. If a cloud is seen inside a building, firemen should not enter except to complete a rescue. An explosion is very likely. A vapor cloud does not necessarily show the limits of the flammable gas, but merely the limits of its refrigeration effect. The flammable gas may extend beyond this on all sides. Therefore, fire fighters must keep low behind their fog pattern and should never enter or closely approach the vapor cloud.

If tanks must be removed, protect personnel with water. If a small, leaking portable tank cannot be shut off and must be moved to a safe place, it should be transported in an upright position so that gas, not liquid, will leak. The tank should never be dragged, for this can damage valves and piping, possibly increasing the flow. Righting an overturned tank should be done carefully. Above all, keep a spray stream on the tank being moved. Portable containers exposed to heat should be taken to a safe place, but *consider carefully before you move a tank on fire*. It is fairly safe while burning if a cooling stream is kept upon it.

Water will protect tanks and exposures. If escaping gas is on fire, immediately apply large quantities of water to all surfaces exposed to heat. Heavy-stream appliances are very desirable and should be applied to all containers, piping, vessels, exposed tanks, and combustible surfaces. The discharge from burning relief valves can create a giant torch seriously endangering not only exposures, but also the tank itself. If it impinges upon the container, there is a possibility of an explosive rupture. Hose streams will cool metal exposures, preventing possible explosive ignitions or ruptures of tanks, and may even lower temperatures and lessen pressures enough for spring-loaded valves to close and cut off the torch effect.

If there is too much heat for the amount of water being applied, a bubble or blister may form on the tank, the noise level from a leak suddenly increase, and the size of the torch flame suddenly grow. *This is your signal for an immediate withdrawal.*

Two Case Histories

Two fires involving LP gases are very instructive. One occurred in a large propane tank farm; the other, on a western highway.

The first is taken from an AIA report.

> The Chief and Deputy Chief of the Newark Fire Department were making a routine inspection of the Warren Petroleum Company on July 7, 1951 when a fire started in a group of propane tanks. Three minutes after the fire was first observed, a ball of fire mushroomed high into the air. Ten or fifteen minutes later, there was a violent explosion due to failure of a propane storage tank. For the next hour and forty minutes, tanks erupted at intervals. Intense radiated heat, which developed as a result of the blasts, caused the fire companies to withdraw their apparatus and personnel to safe distances and started several fires in the surrounding area.
>
> Seventy propane tanks were destroyed or badly damaged. The majority of the tanks were open longitudinally; a number opened up circumferentially. Most of the tanks that appeared to have been subjected to local overheating, yielded, stretched to a minimum

> thickness, and then finally ruptured. A check of the tanks showed that a large percentage of them ruptured along the upper part of the tank lengthwise, along the vapor space.
>
> Many, as a result of rocket effect, traveled distances from one-quarter to one-half mile. A Texaco service station, half a mile from the scene of the blast, was demolished by a falling propane tank. Other tanks ripped up the ground. All the ruptured tanks discharged propane that, in turn, furnished additional heat to rupture other tanks. Railroad rails were twisted and bent. One tank, skyrocketing into the air, returned, and drove itself into the ground, rupturing the water main that supplied water for fire protection.
>
> Two hours after the original fire began the explosions subsided and firemen were able to enter the area to combat the fire. The breaking of the water main forced the fire department to resort to hose relay operations. Firemen used hose streams on tanks to keep the tank surface next to the vapor space cool, but allowed the fire to burn itself out.
>
> The increase in metal temperature within the vapor space of some tanks was so rapid that the means provided for pressure relief could not function rapidly enough to prevent rupture. Yet, pressure relief capacity more than met the requirements of NFPA Pamphlet 58. The three end tanks of a 30-tank section caught fire during the explosions. The fire department placed hose streams on these tanks. In spite of the fact that one tank blistered and finally split open, the water was able to keep the contents sufficiently cool to prevent the tank from leaving its foundation. This demonstrates the effectiveness of water for this purpose.

The second fire illustrates a common misconception about the use of water on LP gas tanks.

> A tank truck and trailer combination, carrying liquefied propane, overturned on a California highway. Fire was immediate and ignited close exposures. Four different fire departments responded and called for advice from an oil company. When representatives arrived, they found that the exposure fires had been extinguished but that several spots on the tank truck and trailer were still burning freely. The oil men urged that cooling water be applied to prevent rupture of the containers. They were told that the firemen had been previously advised not to do so, that the shock of cold water on the hot steel tank would supposedly cause a fracture. The oil company representatives finally convinced the firemen there was little danger of this. On the contrary, if water was not applied quickly, far more serious results could be expected.

Water was then applied to the tanks and, in a short time, the relief valve closed. The remaining propane was consumed by extending vent pipes away from the truck and trailer to a controlled burning. (While controlled burning is an accepted practice under the right circumstances, this does *not* mean you should attempt to ignite a vapor cloud. This is a time when competent technical advice can be invaluable.)

Both the LP gases are nontoxic, but they will cause drowsiness in high concentrations, or produce nausea, headache, or possible asphyxiation. These effects can be avoided by the use of self-contained gas masks.

Refrigeration

The principle that causes the air-cooling effect of an expanding LP gas is put to work in refrigeration systems. In the area to be refrigerated, a liquid is allowed to expand into a gas. This expansion takes up heat from the environment. Low-pressure gas, with a high heat content, then leaves the refrigerated space and moves to a compressor. As a high-pressure gas, with a high heat content, it flows from the compressor to a condenser. Here, the heat which keeps it a gas is given off either to the outside air or to a cooling agent of some kind. As it loses heat, the refrigerant condenses into a room-temperature liquefied gas, with a low heat content, that flows back to a receiver and through an expansion valve. Once again, it is allowed to evaporate in the refrigerated space. In this cyclic compression and expansion, the refrigerant transports heat from the area to be cooled to a designed place of discharge. (This is an extremely simplified version of one kind of system. There are many complex variations on this theme.) For a diagram of the system described above, see Figure 6–2. Some common locations of refrigeration systems are:

1. Household refrigerators and freezers.
2. Commercial refrigeration in butcher shops, frozen food cases, taverns, restaurants.
3. Air conditioning of all types.
4. Industrial refrigeration in meat packing plants, produce storage houses, creameries.
5. Refrigerated transportation in trucks, trains, and ships.

Freons

From a fire fighter's standpoint, the most important variation in these systems lies in the hazards of the refrigerant used. With the exception of ammonia in certain industrial occupancies, and some liquefied nitrogen

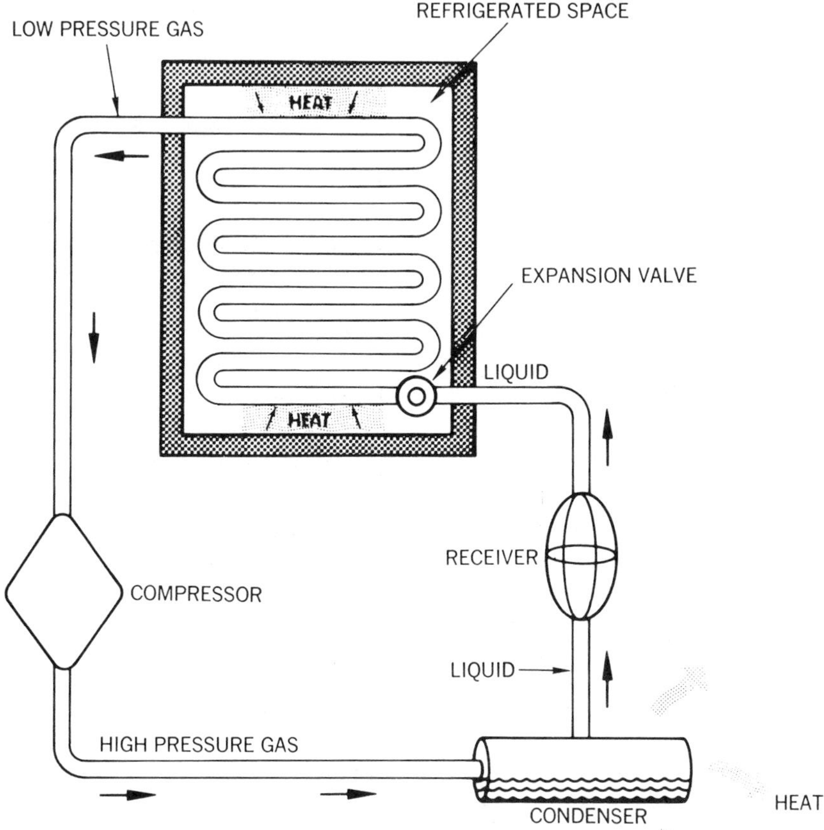

FIGURE 6–2. Refrigeration Cycle

being used in refrigerated trucks and trailers, practically every refrigeration system being built today uses some form of Freon.

Each refrigerant requires a different amount of compression. Some will produce far lower temperatures in the refrigerated space, and each has a different pressure-temperature relationship. For these and other reasons, there are many kinds of refrigerants instead of only one. Several Freons are currently in use, the most common being Freon-12 (dichloro-difluoromethane.)* A summary of the properties of seven Freons is included in Table 6–2 (page 134).

---

*Freon, a brand name of the DuPont Company, refers to halogenated hydrocarbons which contain fluorine. When one or more of the hydrogen atoms in a hydrocarbon is replaced by a halogen (fluorine, chlorine, bromine, or iodine), a halogenated hydrocarbon (halocarbon) results. The jaw-cracking names of the Freons merely indicate what halogens in what number (mono, di, tri, tetra) have hooked on to the carbon atom or atoms.

TABLE 6–2. SUMMARY OF PROPERTIES OF SEVEN FREONS

REFRIGERANT	BOILING POINT	VAPOR DENSITY	TEMP.-PRESS. RELATIONSHIP*	UL TOXICITY GROUP**	COMMENTS
Freon-11 (CCl_3F)	75° F. 24°C.	4.7	227 (108)	5	(Trichloromonofluoromethane) Nonflammable
Freon-12 (CCl_2F_2)	−22° F. −30°C.	4.2	108 (42)	6	(Dichlorodifluoromethane) Nonflammable
Freon-13 ($CClF_3$)	−115° F. −82°C.	3.6	−2 (−19)	6	(Monochlorotrifluoromethane) Nonflammable
Freon-21 ($CHCl_2F$)	48° F. 9°C.	3.5	189 (97)	5	(Dichloromonofluoromethane) Very weakly flammable, ignites at 1026°F.
Freon-22 ($CHClF_2$)	−42° F. −41°C.	3.0	75 (24)	5	(Monochlorodifluoromethane) Very weakly flammable, ignites at 1170°F.
Freon-113 ($C_2Cl_3F_3$)	118° F. 48°C.	6.4	280 (138)	4–5	(Trichlorotrifluoroethane) Very weakly flammable, ignites at 1256°F.
Freon-114 ($C_2Cl_2F_4$)	38° F. 3°C.	5.9	180 (82)	6	(Dichlorotetrafluoroethane) Nonflammable

*This is the temperature in degrees Fahrenheit (and Celsius) required for the gas to reach 10 atmospheres pressure, approximately 147 psi.

**Underwriters' Laboratories Toxicity Classifications:

1: Concentrations of ½% to 1% can cause death or serious injury in 5 min.
2: Concentrations of ½% to 1% can cause death or serious injury in 30 min.
3: Concentrations of 2% to 2½% can cause death or serious injury in 1 hour.
4: Concentrations of 2% to 2½% can cause death or serious injury in 2 hours.
5: Hazards are intermediate between Group 4 and Group 6.
6: Concentrations of 20% are not injurious after 2 hours.

A practical knowledge of the types of refrigerants presently being used can be of great value in handling fires, explosions, or gas leaks involving them.

Explosions can be caused by overpressure of a gas within a system. To prevent these explosions, especially during a fire, refrigerant pressure vessels are equipped with pressure relief devices large enough to relieve a rise in pressure even though the system is fully involved in a hot fire. Some smaller Freon systems may use fusible plugs that will only protect them against rising temperatures. But there are other causes of overpressures besides temperature rise. An example is the clogging of a frozen part as the compressor forces in more refrigerant. For these reasons, pressure relief devices are preferred to fusible plugs. All gases are subject to pressure explosions if they are confined and heated, even if, like the Freons, they are almost totally nonflammable.

Toxicity

All the Freons have a low order of toxicity. There is a misconception that the Freons will produce quantities of phosgene gas when exposed to fire temperatures. This probably comes from assuming that the Freons are similar to carbon tetrachloride. Freons do *not* produce significant amounts of phosgene. However, they do produce halogen acid gases like hydrogen chloride. This means that protective equipment and self-contained gas masks are required for fire fighters exposed to heated Freons.

When a concentration of 5 percent Freon or more, in air, surrounds large oil or wood fires with no ventilation, it can yield lethal concentrations of acid gas. Acid gases are less dangerous than phosgene because they are very irritating to the nose and throat and therefore easily detected in very small amounts. It is almost impossible to remain voluntarily in an atmosphere containing toxic percentages. This is not true with phosgene. Halogen acid gases from burnt Freons can damage metal and clothing. Some situations where this has been observed include the discharge of Freons into the air while a clothes dryer, a stove, or an oven were in operation. In these cases, damage to clothing and metal was evident. With this in mind, it should be remembered that Freons are often used as propellants in aerosol cans.

Refrigerant Fires

If refrigeration is involved in a fire, first determine what refrigerant is being used. If it is Freon, pull the main electrical switches to stop the compressors. For air conditioning systems, stop the fans if possible to prevent the spread of fire. *Do not dump the refrigerant.* If the refrigeration piping and vessels are in the fire area, provide ventilation for the burnt Freon acid gases.

On a large system, if Freon is escaping from damaged lines or equipment, try to turn off the main liquid supply valves from the refrigerant receivers. *No harm can come from turning off any or all valves as long as the compressor switch is off.* Maintain a check for leaking refrigerant. As soon as possible, call the refrigeration service company for technical advice. Valves may have to be closed to protect the system against contamination from air or moisture. Perishable products may be in danger of spoilage from rising temperatures. One of the major causes of accidents in refrigeration fires is ice on the floors. Keep water seepage into cold storage rooms to a minimum.

A Freon leak can be detected with the use of a halide lamp. The flame color of the lamp will change to blue or bright green if a small quantity of Freon is present. *Do not use this lamp around flammable gases or other refrigerants.*

Ammonia

A brief description of ammonia in a chemical dictionary could look like this:

> Anhydrous ammonia, NH_3, a colorless gas with an extremely pungent odor, is highly irritating to the eyes, skin, and respiratory tract. It is flammable, with an ignition temperature of 1204°F. (651°C.) and a flammable range of 16 to 25 percent. The gas is extremely water soluble and has a vapor density of 0.6. Anhydrous ammonia is shipped in tank cars, tank trucks, and steel cylinders with a DOT green label required. It is widely used in many industrial processes, as a fertilizer and refrigerant.

While these are the necessary bare bones, the hazards of ammonia cannot be appreciated until some flesh is added. The very first word in our brief description requires some explanation: *Anhydrous* means "having no water." At normal temperatures and pressures, anhydrous ammonia is a dry gas with a boiling point of –28°F. (–33°C.). Since its critical temperature is 271°F. (133°C.), it can be liquefied by pressure into liquid anhydrous ammonia. Liquid anhydrous is stored like any other liquefied gas in tanks of various sizes, and has its own pressure-temperature relationship. (See Figure 6–3.)

Never confuse liquid anhydrous ammonia with household ammonia, which is something else. When ammonia gas is dissolved in water it becomes ammonium hydroxide or aqua ammonia. While it is true that some ammonium hydroxide solutions, containing 30 percent ammonia gas in water, are much stronger than household ammonia, they are all ammonium hydroxide and less dangerous than liquid anhydrous ammonia.

To summarize, there are three different forms of ammonia. All of them are called ammonia. Do not mistake one for the other.

1. Anhydrous ammonia is the pure dry gas.
2. Liquid anhydrous ammonia is this gas compressed into a liquid.
3. Ammonium hydroxide is gaseous ammonia dissolved in water. (Household ammonia is a weak ammonium hydroxide solution.)

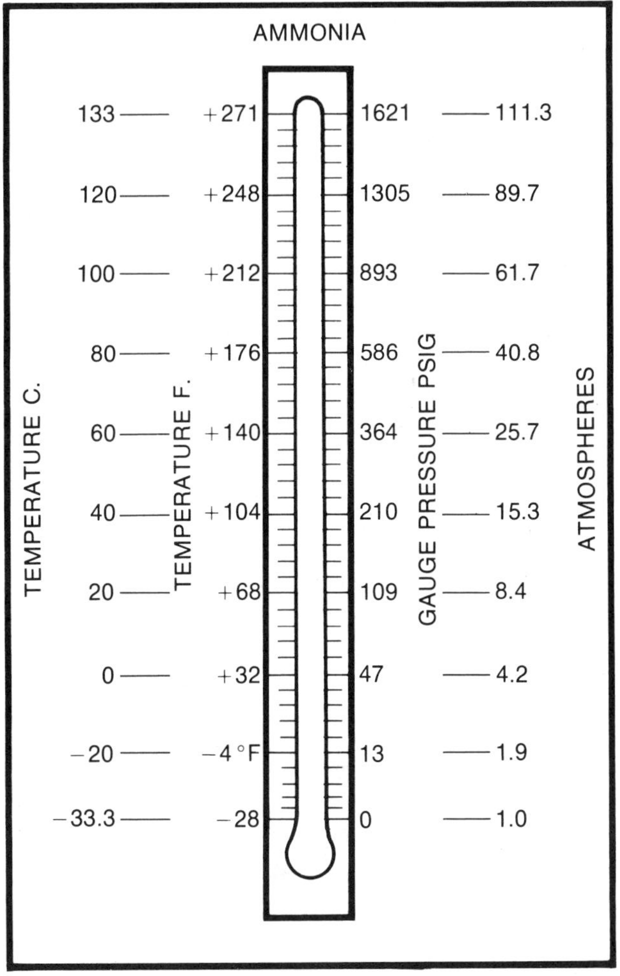

FIGURE 6–3. Temperature-Pressure Relationship of Ammonia

Ammonia Hazards

Anhydrous ammonia is flammable but its flammable range is extremely high: a 16 percent concentration in air is needed just to reach the lower flammable limit. This is why the DOT classifies ammonia as a nonflammable gas, and only requires a green label. This label has caused some furor because ammonia fires and explosions are not uncommon. Once again, a small amount of liquid will generate a great deal of gas, even enough to reach a lower flammable limit of 16 percent. Yet, even if liquid anhydrous is not present, ammonia gas is explosive. It is even being considered as a possible rocket fuel. As we shall see, the presence of contaminants can greatly increase its hazard.

Nevertheless, anhydrous ammonia doesn't approach the fire hazards of such liquefied gases as propane or butane. Not only is the flammable range higher, but ammonia is lighter than air. Unlike the LP gases, it will diffuse rapidly in an open area. In addition, anhydrous ammonia is extremely water soluble. A sufficient amount of water fog can rapidly turn an explosive concentration of gas into a relatively harmless ammonium hydroxide solution. Ammonia fires are fought like any other flammable gas; the primary object is to halt the gas flow.

In this regard, an ammonia leak or fire can be very difficult to bring under control. It can be a problem to reach and shut off valves because ammonia is far more toxic than the hydrocarbon gases and personnel must take extra precautions to protect themselves Fortunately, the piercing odor of ammonia is unmistakable and unforgettable. Anyone not trapped or injured will immediately vacate the area. (This is one case where a wet handkerchief or towel held over the nose can give someone trying to escape temporary relief.) Low concentrations of ammonia gas can severely injure the respiratory membranes with fatal results. It causes the throat passages to swell, blocking the airways. The Underwriters' Laboratory (UL) considers it a Group 2 toxic hazard: a concentration of 0.5 percent in air will produce serious injury or death in 30 minutes. All fire fighters should be equipped with self-contained gas masks. Canister masks, even those specially designed for ammonia, will only filter out limited percentages.

The liquid-vapor ratio of liquid anhydrous must again be considered, for high percentages of ammonia gas can suddenly appear. Moderate concentrations of ammonia gas can cause fire fighters, even if they are equipped with respiratory protection, painful irritation of tender skin at the back of the neck, the forehead above the mask, the insides of the wrists, and the genital area. Gas-tight vapor suits are the only absolute protection, although prompt ventilation and the use of water fog can appreciably lessen the hazard. If desired, water fog will also extinguish ammonia fires. So will dry chemical and CO_2. Liquid anhydrous is even more irritating than the gas and causes severe skin burns on contact. Ammonium hydroxide, especially in the higher strengths, will also cause burns, and its vapors should be avoided by the use of self-contained masks.

Ammonia Storage

Liquid anhydrous ammonia is shipped in a wide range of containers under the controversial DOT green label. DOT-approved cylinders are generally required to have a pressure relief device to prevent overpressure explosions. In the past few chapters, we have mentioned three types.

1. Frangible discs that burst when overpressures occur for any reason.

2. Fusible plugs that melt at designed temperatures, generally somewhere between 157°F. and 220°F. (70°C. and 104°C.). These only protect the cylinder against overpressures caused by heat. (Some protective devices combine frangible discs with fusible plugs to prevent bursting at normal temperatures.)

3. Pressure relief valves that open at a predetermined pressure or heat and close again when this pressure or heat is relieved.

The DOT has several exceptions to its pressure relief device requirement. Cylinders containing many types of nonliquefied gases stored at less than 300 psi (20 atmospheres) need none. Some, but not all, poisonous gas cylinders need none. Neither do small ammonia cylinders with a capacity of less than 165 pounds. Meanwhile, ammonium hydroxide is shipped in everything from pint bottles to 8000-gallon (30,000-liter) tank cars. Despite its toxicity, no labels are required and there are no shipping regulations.

Liquid anhydrous ammonia tanks should be located in cool, well-ventilated places, preferably outdoors. Large containers are equipped with pressure relief devices. If inside, these tanks should be in a fire-resistive room, underneath a sprinkler system. They should be separated from other chemicals, particularly oxidizing gases, halogens, and acids.

Ammonia is a relatively simple compound of two gases, nitrogen and hydrogen. The molecule is diagrammed in Figure 6–4.

Ammonia decomposes into nitrogen and hydrogen at temperatures between 840°F. and 930°F. (449°C. and 499°C.). It is often broken apart, under controlled conditions, to provide supplies of these gases. Both hydrogen and nitrogen have many industrial applications. Nitrogen is used for protective inert atmospheres, in explosives, in many compounds, and, when ammonia is applied directly beneath the ground, as a fertilizer. We discussed some uses of hydrogen in the last chapter. Ammonia is also valuable on its own, to vulcanize rubber, to extract metal from ores, to manufacture some types of paper, to refine oil, to treat water, and to form

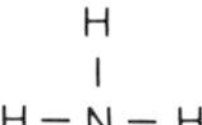

FIGURE 6–4. Ammonia (NH_3)

ammonium hydroxide, which is used in pharmaceuticals, soaps, inks, ceramics, and detergents. Anhydrous ammonia is also a very efficient refrigerant.

An ammonia system can be recognized by its steel pipe. Moist ammonia reacts with copper, zinc, tin, silver, and their alloys, causing corrosion. This is why brass or copper piping, soldered joints, or galvanized iron are never used around ammonia. Ammonia compressors are always driven by completely separate motors through v-belts and couplings. The locations of an ammonia leak can be pinpointed through the use of a sulfurous gas that turns white and becomes easily visible.

Fire Histories

There can be explosions of ammonia either at the start of or during a fire. A 16 percent concentration takes some time to build up—enough time for the fire department to arrive at the scene and to begin work. In Iowa, a fire followed a break in a large ammonia pipeline. Probably, the ammonia became mixed with air and lubricating oil vapors. Employees in other buildings saw the white fumes pouring from windows and sent in a fire alarm. The explosion occurred just as two men were entering the engine room to shut down the generator.

Ammonia explosions can be powerful. Many reports speak of windows broken throughout a large area, collapsed roofs, and buildings blown apart.

The explosions can be multiple, each explosion increasing in severity. In a Miami milk-bottling plant, release of ammonia was caused by failure of a 2-inch nipple on an ammonia compressor. The first explosion was relatively minor but attracted the attention of the relief engineer just after 5:00 A.M. His investigation revealed no flame, but a second explosion, ten minutes later, started a raging fire in the engine room. In another ten minutes a third and still heavier explosion spread the fire throughout the building. The loss was $350,000.

Even without an explosion, the ignition of ammonia can accelerate combustion. One report speaks of escaping ammonia gas that resembled a torch as it mushroomed against the ceiling. In many ammonia fires, the intensity became so great it was difficult or impossible for the fire department to get close enough to make an effective attack. Such phrases as "fully involved" or "70 percent of the building was fully involved upon our arrival" are commonplace in histories. Although the upper flammable limit of ammonia is 25 percent, some reports mention the fact that concentrations of gas in the immediate area of the leak were too rich to burn. Fires and explosions were more severe some distance away from the leak where the ammonia was within its flammable range.

The accidental rupturing of an ammonia vessel with the release of unignited fumes is serious enough. Some reports speak of the panic of employees. We can imagine what happened when ammonia fumes became strong within a building where 70 women were employed in processing vegetables.

In spite of our best efforts, tragedies still occur in ammonia fires: in a Michigan ice cream store, two fire fighters were making their way through flames on the first floor in an attempt to take a hose down the rear stairway when an ammonia tank in the basement exploded. Several other fire fighters were overcome by fumes but managed to reach safety. Similarly, fire fighters wearing masks were making slow headway against a stubborn fire in the basement of a Rhode Island food market when another large ammonia tank exploded. Eleven of them were trapped and overcome.

Ignition Sources

Ignition sources for these ammonia fires and explosions vary widely. The list includes oil burners; pilot lights; electric wiring, broken and arcing due to the force of a tank rupture; light bulbs shattered through contact with cold ammonia vapors; static electricity generated by rapidly escaping gas; a system becoming involved in a fire of another origin; and the closing of electrical switches. In Alaska, ammonia filled a building while an employee was checking the oil in one of the compressors. He was forced out of the building by the fumes before he could shut down the compressor. He then went to the transformer building and pulled the main power switch. As he did so, an immediate explosion occurred that raised the roof of the one-story building three feet and involved the entire building in flames.

This history suggests that the pulling of main electrical switches to stop compressors can be dangerous if there is a flammable concentration of ammonia. This is troublesome. One of our primary objectives, if ammonia is escaping, is to try to turn off the liquid ammonia supply valves. We can be certain that no harm can come from closing any or all valves if the compressors are stopped. Yet, apparently, we dare not pull the main switches to stop the compressors. This is why prompt notification of the refrigeration service company is important. Technical advice may be mandatory. Technicians or engineers of the company in trouble may have been at the scene just before a rupture, a leak, or an explosion. Reports indicate that they are often incapacitated before the fire department arrives, and, to the uninitiated, a refrigeration system can seem a maze of pipes. Sometimes firemen close the wrong valves during emergencies, thus aggravating the problem.

Ammonia Diffusion Systems

Use of a fire department diffusion system can lessen the danger if an ammonia system is in immediate jeopardy from fire. These diffusers will mix the entire charge of ammonia with water and dump it down the sewer. A fire department box on the outside of the building, preferably in an easily accessible and plainly marked location, contains a valve to dump high-pressure gas and liquid, a valve for the low-pressure gas, and a water valve. It also contains a 1½'' or 2½'' inlet for additional fire department water supplies. Since a gallon of water absorbs one pound of ammonia gas, the entire amount of ammonia can eventually be mixed into a safer solution of ammonium hydroxide. But this dumping can take some time. Some systems contain more than ten tons of ammonia. *Start the water flowing before the valves are opened to dump the system.* Each fire department should be familiar with the location and operation of all ammonia diffusers within its city. In addition, most ammonia systems have pressure relief valves located so they discharge into water tanks or to the air. If they open to air, note their location. Keep personnel upwind and sources of ignition away.

Other Refrigerants in Use

Before ending this section on refrigerants, some others should be mentioned briefly. While they are being replaced by the Freons, some of them can still be found in older systems. Despite their flammable hazards, a few hydrocarbons are used: butane (No. 600); ethylene (No. 1150); propane (No. 290); ethane (No. 170); and isobutane (No. 601).* Isobutane has properties similar to butane, an ignition temperature of 864°F. (462°C.), and a flammable range of 1.8 percent to 8.4 percent. Several chlorinated hydrocarbons with varying degrees of flammability and toxicity are still in service. They include dichloroethylene (No. 1130); dichloromethane (No. 30); methyl chloride (No. 40); and ethyl chloride (No. 160).

While carbon dioxide (No. 600) and sulfur dioxide (No. 764) will be covered more fully in the chapter on toxic combustion products in *Explosive and Toxic Hazardous Materials*, an additional word about SULFUR DIOXIDE must be included here. It is considered the most toxic of the refrigerants, the only common one with a UL toxicity classification of 1. In many large cities, it can still be found in the central refrigeration systems of old apartment houses.

Like ammonia, the choking odor of sulfur dioxide gives immediate warning of its presence. It is about twice as heavy as air and will linger awhile. It is not flammable. Sulfur dioxide systems may discharge into tanks containing caustic water and lye solutions. Ammonium hydroxide will react with sulfur dioxide to form a harmless white cloud. Household ammonia, in

---

*These numbers are the designation given to various refrigerants by the American Society of Refrigeration Engineers (ASRE). They represent the chemical formulas numerically. Ammonia is refrigerant No. 717, and the Freons carry their numbers in the names. Freon-12 is refrigerant No. 12. The usefulness of these numbers is limited. Refrigerants are far more often called by their names.

front of blowing smoke ejectors, can sometimes be used when householders become overly enthusiastic in their use of sulfur dioxide fumigants.

METHYL FORMATE (No. 611) is a colorless liquid with an agreeable odor. While it is moderately toxic and irritates the eyes, industrial deaths from this material are extremely rare, occurring only when high concentrations are encountered. Its vital statistics are as follows:

FLASH POINT	IGNITION TEMPERATURE	FLAMMABLE LIMITS	SPECIFIC GRAVITY	VAPOR DENSITY	BOILING POINT	WATER SOLUBLE
−2° F. −19°C.	853° F. 456°C.	5.9–20%	1.0	2.1	90° F. 32°C.	Yes

Finally, a mixture of two gases, Freon-12 and ethylidine fluoride (respectively, 73.8 percent and 26.2 percent) is now being used as a refrigerant. It is sometimes called Freon-500. It is practically nonflammable, has no warning odor, and has a UL toxicity rating of 5. Once mixed, these gases act like a single substance in that the vapor has the same composition as the liquid. Such a mixture is called AZEOTROPIC.

QUESTIONS ON CHAPTER 6

1. What is the difference between liquid anhydrous ammonia and ammonium hydroxide?
2. How may water be used on LP gas fires?
3. What is the difference between critical temperature and critical pressure?
4. When Freons are overheated, what type of gases do they produce?
5. Discuss fire department use of ammonia diffusion system.

BIBLIOGRAPHY

Union Carbide Corp. (Linde Division):

"Precautions and Safe Practices for Handling Atmospheric Gases," 1957;
"Production of Industrial Gases from the Air," 1957. Also, a whole series of pamphlets on individual gases.

NFPA Standard No. 58:

"Storage and Handling of Liquefied Petroleum Gases."

American Insurance Association:

Special Interest Bulletins:

No. 4, revised July 30, 1953, ''Liquefied Petroleum Gases.''

No. 20, revised May 15, 1950, ''Bottled Gas Systems and the Inherent Hazards.''

No. 143, November 1959, ''Liquefied Petroleum Gas Fire Control.''

Scott Aviation Company:

''Gaseous Oxygen Fires and Explosions,'' by Arthur M. Miller. 1961.

Ray Winther Company (South San Francisco):

''Hazards of Refrigeration Systems: Explosions Involving Ammonia,'' 1967.

Fireman's Magazine:

''Things to Consider About LP Gas'' (September 1955).

''LP Gases: What They Are; What They Do'' (April and May 1959).

''How to Handle LP Gas Vehicle Incidents'' (August 1959).

''What About LP Gas Cylinders?'' (October 1965).

VISUAL AIDS

''Handling LP Gas Emergencies,'' produced by the National Fire Protection Association, 24 minutes, color; excellent.

DEMONSTRATIONS

An ammonia pump, available from any well-equipped chemistry laboratory, will demonstrate the water solubility of ammonia. The demonstration should be conducted in the laboratory.

7 Cryogenics

Pressurization is not the only way to liquefy a gas. Cooling any gas below its boiling point will also do the job. A gas is a liquid or a solid according to the relationship between its boiling and freezing points and the temperature. We think of water in its three forms only because normal temperatures in temperate zones produce both the solid and liquid states and a hot summer sun evaporates a puddle before our eyes. Yet, it is not hard to think of places here on earth where ice never melts and others where water is vaporized rapidly. The only reason water is predominantly a liquid is that our normal range of climate is neither too warm nor too cold.

There are many liquids which have boiling points within the temperature range of a hot summer day and a cold winter night. Whether they are gases or liquids depends not only on the month, but even on the time of day or night. Methyl bromide boils at 40°F. (4°C.). It can be liquefied by putting it into a refrigerator. Table 7–1 (pages 146–147) lists some of these indecisive liquids. We must be prepared to deal with them, either as flammable liquids or gases, with fire-fighting measures appropriate for each state.

Many gases, particularly elemental gases, have extremely low boiling points. Their critical temperatures are almost as chilly. They cannot be liquefied at ordinary temperatures by any amount of pressure. How can industry have the convenience of liquefied storage at ordinary temperatures when a gas has a boiling point of −423°F. (−253°C.)? Methods to reduce their temperature to this subantarctic range had to be found, and, once found, a technology had to be developed to keep them liquefied although they are stored at temperatures 500 degrees F. (278°C.) above their boiling points.

TABLE 7–1. BOILING POINTS OF SELECTED FLAMMABLE LIQUIDS

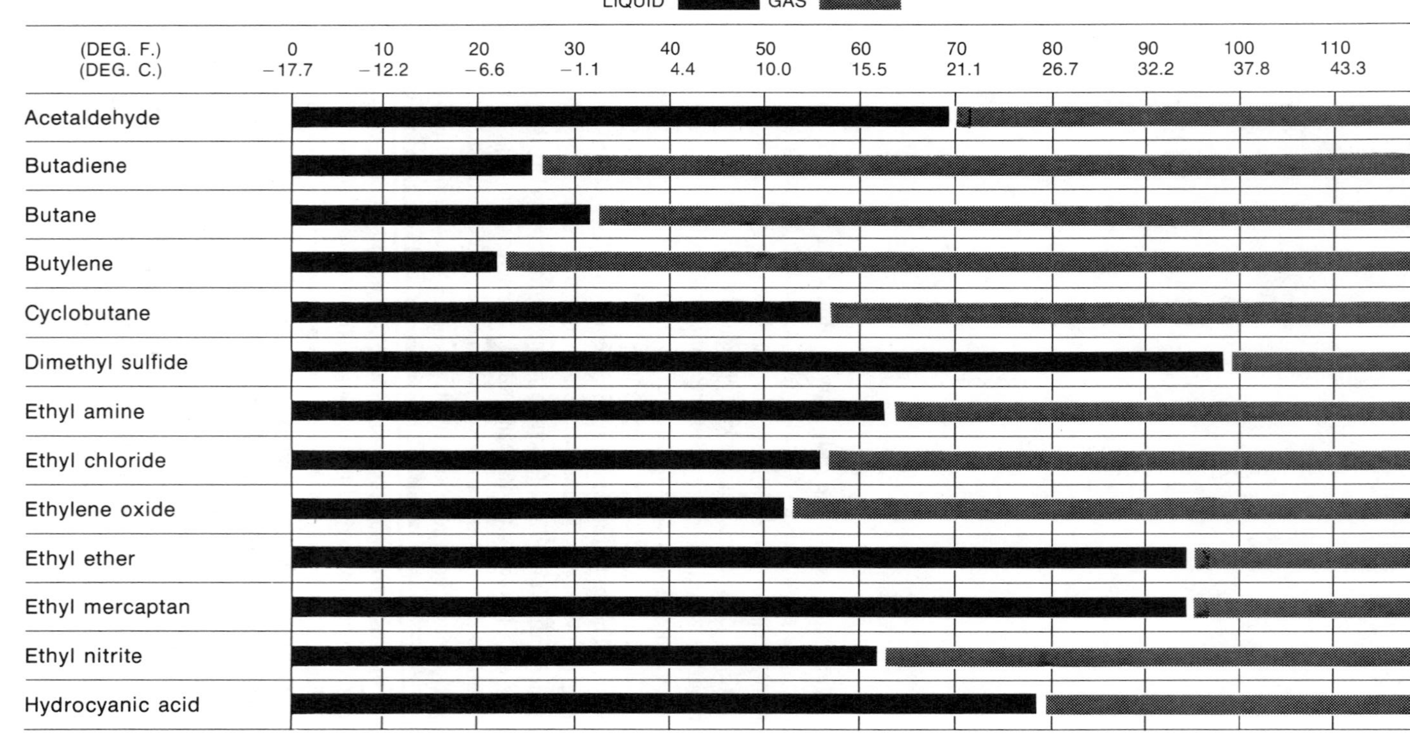

TABLE 7–1. BOILING POINTS OF SELECTED FLAMMABLE LIQUIDS (CONT.)

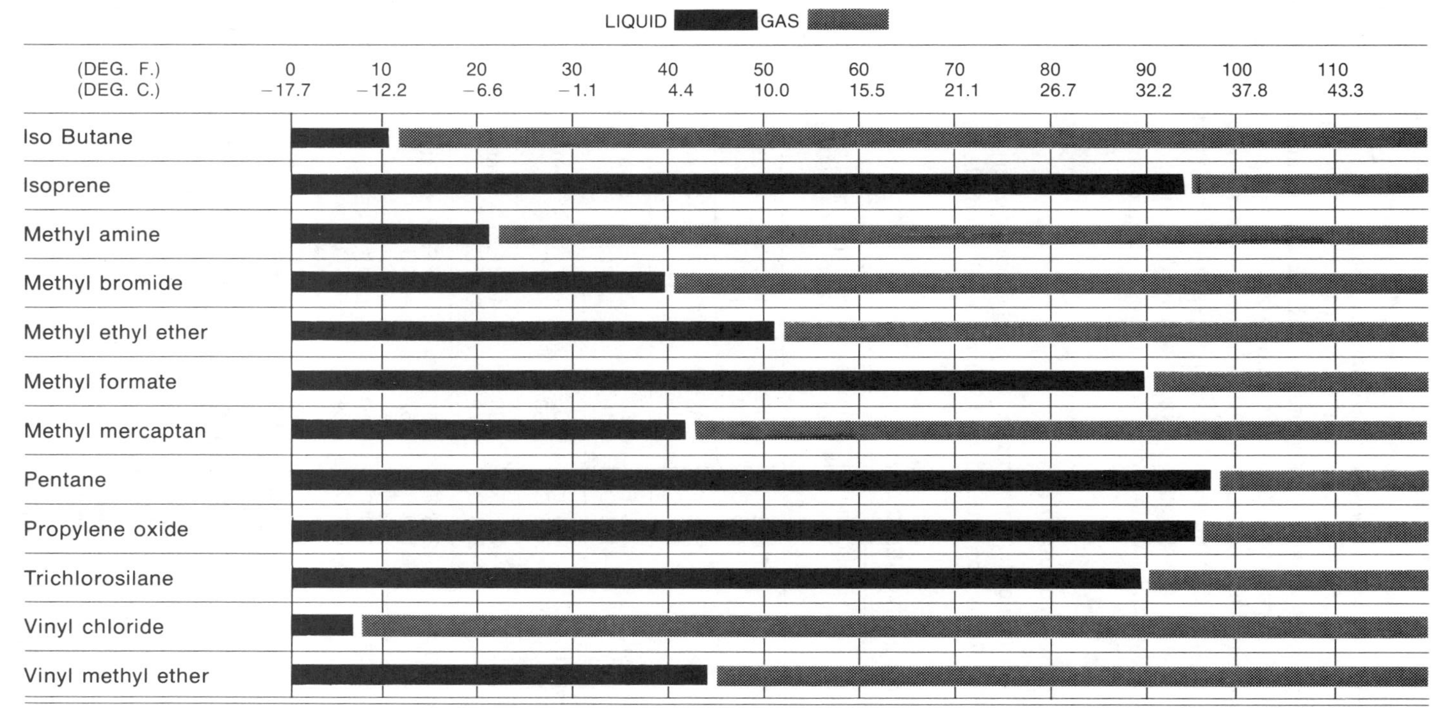

The answers have been a long time in coming, but a new science is solving many of the problems. The science of CRYOGENICS is defined as *the production and use of materials at temperatures ranging from* −150°F. (−101°C.) to absolute zero, −459.7°F. (−273°C.). Producing cryogenic gases, gases at ultralow temperatures, has become an industry with enormous growth potential. One economic reason for the preference given to liquefied gases can be found in the liquid-to-gas ratio shown in Table 7–2. For example, 862 volumes of gaseous oxygen can be liquefied down to a single volume. A cryogenic cylinder can hold 12 times more gas than a pressurized cylinder of the same size. Cryogenic gases are good business.

Cryogenic Production

The production of liquid nitrogen is a good example of cryogenic production methods. Air is compressed to 1500 psi (102 atmospheres). As we have

TABLE 7–2. CRYOGENIC GASES

	BOILING POINT	MELTING POINT	CRITICAL TEMP.	CRITICAL PRESSURE	LIQUID-TO-GAS RATIO (VOLUMES)
Argon (Ar)	−302° F. −186°C.	−308° F. −189°C.	−188° F. −122°C.	48 atm.	846 to 1
Air (Mixture)	−318° F. −194°C.	−363° F. −219°C.	——	37 atm.	728 to 1
Fluorine (F_2)	−306° F. −188°C.	−363° F. −219°C.	−200° F. −129°C.	55 atm.	981 to 1
Hydrogen (H_2)	−423° F. −253°C.	−435° F. −259°C.	−400° F. −240°C.	13 atm.	840 to 1
Helium (He)	−452° F. −269°C.	−438° F.* −272°C.	−450° F. −268°C.	2 atm.	755 to 1
Krypton (Kr)	−243° F. −153°C.	−251° F. −157°C.	−88° F. −67°C.	54 atm.	694 to 1
Methane (CH_4)	−289° F. −178°C.	−299° F −184°C.	−116° F. −82°C.	45.7 atm.	637 to 1
Neon (Ne)	−411° F. −246°C.	−416° F. −249°C.	−380° F. −229°C.	27 atm.	1445 to 1
Nitrogen (N_2)	−320° F. −196°C.	−346° F. −210°C.	−233° F. −147°C.	33.5 atm.	697 to 1
Oxygen (O_2)	−297° F. −183°C.	−361° F. −218°C.	−181° F. −118°C.	50 atm.	862 to 1
Xenon (Xe)	−163° F. −108°C.	−169° F. −112°C.	62° F. 17°C.	58 atm.	559 to 1

*at 26 atmospheres pressure

learned, when a gas is compressed, it produces heat. This heat of compression is dissipated in several ways, most cheaply by the use of cold water. We now have compressed air at 1500 psi (102 atmospheres) at a temperature of 32°F. (0°C.). Additional compression raises the pressure to 2000 psi (136 atmospheres). The small amount of additional heat is again dissipated. The compressed air is then cooled to −40°F. (−40°C.) by an ammonia refrigeration system. The air is next led to a larger tank and allowed to expand. The expansion of a gas cools it. As the pressure drops from 2000 psi to 90 psi (136 atmospheres to 6 atmospheres), so does the temperature, from −40°F. to −265°F. (−40°C. to −165°C.).

At this stage, several alternatives are possible: The remaining pressure can be released and the air will drop below its boiling point of −318°F. (−194°C.). If desired, previously produced liquid nitrogen can be used to cool the air to −320°F. (−195°C.) before the remaining pressure is released. In this way, the air's temperature can be driven below −400°F. (−240°C.). Similarly, compressed hydrogen gas, cooled to −400°F. (−240°C.) before its pressure is released, will liquefy at −423°F. (−253°C.); helium gas, cooled by liquid hydrogen, can be liquefied at −452°F. (−269°C.), only eight degrees above the coldest possible temperature.

Once liquid air is formed, it can be allowed to boil off. Since the different gases in air boil at different temperatures, as each forms it can be collected. Or, liquid air can be stored and its various liquid fractions separated according to their specific gravity. For a long time, the most useful and valuable fraction was liquid oxygen. But air is 78 percent nitrogen, which was going to waste. Someone soon found a use for it. In fact, the multiple uses found for liquid nitrogen are a good example of why production of cryogenic gases is constantly expanding.

Uses for Cryogenics

Besides the savings in storage space and weight they offer, cryogenic gases are valuable simply because they are so cold. Liquid nitrogen is used to freeze liquids for emergency pipeline repairs; to solidify gummy materials, such as certain plastics and cosmetics, before they are ground; to preserve human tissue in banks and the spermatazoa of prize bulls before artificial insemination; for shrink fitting of valve inserts, propeller shafts, and bearings; in medicine, for the removal of skin blemishes and for the correction of Parkinson's Disease (liquid nitrogen is released deep inside the brain). This listing is incomplete, and new uses, such as the recent notoriety given the preservation of dead bodies, are found every day.

The use of liquid nitrogen to refrigerate trucks is increasing. A tank inside the storage space expels liquid nitrogen upon command from a thermostat. While nitrogen will combine with several burning metals and is not considered totally inert for that reason, the gas is also used to provide completely safe atmospheres above and around various industrial operations. There would be a small likelihood of any fire inside a trailer when a nitrogen atmosphere is present. If you have to enter the trailer for some reason, remember that nitrogen can act as an asphyxiant. Unless you are wearing a self-contained mask, open the door and wait three minutes. Look out for metal pipes inside. They will be intensely cold, cold enough to "burn" your skin.

Cryogenic gases are not new. The increased use of liquid hydrogen and oxygen in the space program has simply made them more glamorous. Liquid oxygen, nitrogen, and argon have been used, stored, and shipped in large quantities for more than a quarter of a century.

In spite of the fact that we, as fire fighters, may be unfamiliar with the strange properties and hazards of cryogenic materials, we must learn how to handle them in an emergency. It is expected of us.

In a bulletin directed toward police, concerning highway accidents, the American Insurance Association concluded a section on low temperature gases with these words: "The proper authorities, such as the fire department and company officials, should be notified and consulted for guidance as how to best control the particular situation at hand."

Hazards of Cryogenic Gases

Basically, the hazards of cryogenic gases can be reduced to three categories:

1. *The hazards of the particular gas in question,* although the danger may be accentuated in a cryogenic fluid. Liquid hydrogen and methane are still flammable. Liquid oxygen will support combustion with a roar. Liquid fluorine is reactive and toxic.*
2. *The tremendous liquid-to-vapor ratio.*
3. *The extreme cold.*

Since all cryogenic gases have a high vapor ratio and are a freezing threat to a brass monkey, we will concentrate on these two hazards first. Because they are so cold, all cryogenic gases are a danger to flesh. They will "burn" the skin. When spilled on your skin in a small amount, they tend to skitter like a water droplet on a hot stove. In large amounts, they can cling because of their low surface tension. Since these materials are as far removed from your skin temperatures as a hot stove, they can burn you just as badly.

The gases produced by these liquids are also extremely cold. An exposure too brief to injure your hands or skin can damage your eyes by freezing the

*Liquid oxygen, fluorine, and hydrogen are discussed in this chapter and also from a somewhat different perspective in *Explosive and Toxic Hazardous Materials*.

liquid (largely water) in them. For safety, wear goggles, loose asbestos gloves which can be thrown off in a hurry, and trousers with no cuffs. If any liquefied gas contacts your skin or eyes, immediately flood the area with large quantities of unheated water and apply cold compresses. If the skin is blistered or the eyes affected, get to a physician immediately.

All of these gases, except oxygen, are asphyxiating or toxic in themselves. There is also the possibility that a gas with a lower boiling point than oxygen will extract the oxygen from any air exposed to it, reducing the concentration to levels which will not support life.

In a pamphlet, "Precautions and Safe Practices for Handling Liquefied Atmospheric Gases," Linde Division of Union Carbide Corp., one of the leading producers of cryogenic gases, says:

> Liquid nitrogen is colder than liquid oxygen. Therefore, if it is exposed to the air, oxygen from the air may condense into the liquid nitrogen. If this is allowed to continue for any length of time, the oxygen content of the liquid nitrogen may become appreciable and the liquid will require the same precautions in handling as liquid oxygen. However, most liquid nitrogen containers are entirely closed except for a small neck area and the nitrogen gas issuing from the surface of the liquid forms a barrier which keeps air away from the liquid and prevents contamination.
>
> [But] if enough nitrogen gas evaporates from the liquid in an unventilated space, the percentage of oxygen in the air may become dangerously low. When the oxygen concentration in the air is sufficiently low, a man can become unconscious without sensing any warning symptoms, such as dizziness. If a person seems to become groggy or loses consciousness, get him to a well-ventilated area immediately. If you have any doubt about the amount of oxygen in a room, ventilate the room completely before entering it.

Cryogenic liquids will refrigerate the moisture of the air and create a telltale visible fog except when the air is very dry. Unlike the LP gases, this fog normally extends over an area *larger* than the area containing an appreciable amount of released gas. A hydrogen cloud can be several times larger than an ignitable concentration of gas. Reports show that the percentage of oxygen on the edge of a liquid oxygen fog is very close to that of normal air.

Storage of Cryogenics

Because of cold and the vapor ratio, cryogenic liquids cannot be stored in just any container. Their tanks are double-walled, gigantic vacuum bottles. An insulating material, Perlite, or a multi-layered, crinkled, aluminized Mylar film under high vacuum fills the space between the inner and outer jackets. The inner tanks are built of corrosion-resistant stainless steel, aluminum, or various copper alloys. Carbon steel becomes too brittle at these temperatures. Tanks come in varying shapes: vertical, spherical, and cylindrical, and in sizes ranging up to 175 thousand barrels (28,000 cubic meters).

Heat from the atmosphere will penetrate even a well-insulated tank. The temperature differential may be over 500°F. (269°C.). Some of the liquid will absorb this heat and vaporize, but most of the fluid remains at the same temperature. Through this process of self-insulation, the temperature inside a cryogenic container remains surprisingly uniform. So do the vapor pressures. During normal operation, the vapor pressure inside a cylinder will not exceed 100 psi (6.8 atmospheres). But if the tank is unused, more and more gas is formed as heat leaks in. The pressure inside the container will rise slowly.

Pressure Relief Systems

1. A PRESSURE RELIEF VALVE which opens well below any pressure which would threaten the integrity of the tank itself. If the tank is not used for four or five days this valve will open and release the overpressure.

2. A FRANGIBLE DISC which ruptures if the pressure risk is excessive or if the pressure relief valve sticks for any reason. This disc is set above the liquid level. It should only vent gas if the container is upright. The frangible disc will quickly lower the interior pressure of the tank to atmospheric levels. It will operate when the insulation is damaged and the inner tank exposed to normal temperatures. In such a case, the evaporation rate may exceed the capacity of the relief valve.

3. The INSULATION SPACE between the inner and outer jackets is protected by a relief device which functions at a few pounds pressure. This guards against leakage from the inner tank. (Lack of such a device may have been the cause of at least one major cryogenic catastrophe.)

In a test, an intact cryogenic container full of liquid nitrogen, protected by these devices, withstood fire test temperatures of 1800°F. (982°C.) for an hour. Not even the frangible discs blew, since the relief valves handled the pressure rise. At the end of an hour at 1800°F. (982°C.), half the original liquid nitrogen remained, still at −320°F. (−196°C.). For this reason, if we decide that a protective water spray is necessary to help prevent pressure buildups, we must be certain not to jam the pressure relief valve with ice.

If cryogenic fluids are trapped anywhere, at any time, they can cause a violent pressure explosion. Given the chance to expand, their vapor ratios make them menacing. This is particularly true in piping. Pressure relief valves should be installed on every length of pipe between two shutoff valves. All pipes should be sloped up from the tank to avoid the possibility of trapped liquid. Pipes should be made of stainless steel, aluminum, copper, Monel, bronze, or even a plastic which will shatter after a small pressure increase.

Cryogenics Which Don't Support Combustion

While all cryogenic vessels exposed to a fire or entangled in an accident should be approached with due caution, several of the elemental cryogenic gases listed in Table 7–2 are similar to nitrogen in that they will neither burn nor support combusion. They are argon, helium, krypton, neon, and xenon. Once called the ''noble'' gases, these elements used to be considered totally inert and incapable of forming compounds under any circumstances. Recently it was discovered that some of them will combine with fluorine. They have lost their nobility.

ARGON is the most common of the inert gases, making up just less than one percent of the air. Produced in fairly large quantities when air is liquefied, it is used for processing such metals as titanium and zirconium, to fill electron tubes and light bulbs (1.5 billion per year at last count), and to provide inert gas shields.

HELIUM has several interesting uses besides the filling of dirigibles and balloons. Special helium-oxygen mixtures are provided when men must work in high-gas-pressure situations in tunnels or when diving below certain depths. These mixtures help prevent the ''bends,'' caused by the formation of nitrogen bubbles within the body. Helium boils at a lower temperature than any other material. It cannot be solidified at any temperature, even absolute zero, unless pressure is put upon it. More and more liquid helium is being stored and used even though the extremely small difference between its boiling point (−452°F. or −269°C.) and its critical temperature (−450.2°F. or −268°C.) makes this difficult. Strange things happen when helium is cooled to temperatures close to absolute zero. There is no friction, no viscosity, no surface tension; there is a boiling without bubbles, and what seems a repeal of the law of gravity. Some metals immersed in it become super-conductors—that is, once started, an electric current never stops.

NEON, KRYPTON, and XENON are all rare and quite expensive. Seldom, if ever, will you encounter them in large quantities. When an electric arc is passed through neon it glows red; a mixture of neon, argon, and mercury will glow in various shades of blue and green, while helium glows yellow and white.

Liquid Oxygen (LOX)

Even in liquid form, oxygen still will not burn. Somehow this fact never filtered through to the newspapermen who wrote the first accounts of the Apollo tragedy, complete with details about the "burning oxygen" which enveloped the inside of the space capsule. Only when more responsible reports were published were we able to learn the materials whose burning rates were so fatally accelerated by the presence of a pure oxygen atmosphere.

Whenever pure oxygen is present, the sudden ignition and violent burning of combustible materials is possible. The higher the pressure, the greater the hazard. Around pressurized oxygen cylinders, care must be devoted to preventing the sides of the fire rectangle from coming together. Such materials as dust, graphite, oil, or grease should never be allowed to contaminate oxygen valves. Cylinders should never be overheated and valves should be opened slowly to prevent sudden compression (adiabatic heating), which will create heat as surely as the vibration of a loose gasket or the friction of a high-velocity leak.

Whatever can be said about gaseous oxygen supporting combustion goes double in spades for liquid oxygen. While LOX is not flammable either, it enormously increases the burning rate for all combustible materials it contacts. Many substances not normally considered flammable will ignite readily; moderately flammable materials will burn violently; others will explode if mixed with LOX. These potential explosives include all the common hydrocarbons: oil, grease, kerosene, gasoline, acetylene, methane, other flammable gases, tar, dirt contaminated by oil or grease, and even asphalt. There have been reports of liquid oxygen spills which exploded when a fire department hose butt was dropped upon the soaked asphalt of the street. This is why LOX storage should be located on a clean concrete pad.

When such materials as wood, paint, powdered or shredded metals, and cloth are exposed to liquid oxygen, they will also burn violently, or possibly explode, if an ignition source is present. This can happen several minutes after contact. Clothing, soaked or splashed with LOX, should be removed immediately and aired for at least an hour away from all sources of ignition. Fire fighters in uniform or turnout clothing should be relieved to do this immediately.

No open flames or smoking should be allowed near LOX storage. Burning cigarettes dropped into LOX send up a flame two feet long. In the event of a leak, apparatus and personnel should not be allowed downwind. Pure oxygen can have a destructive effect on a gasoline engine. Everyone not actively engaged in fighting a fire around liquid oxygen should be evacuated. The presence of a fire means both heat and a fuel are present. LOX can escalate the situation into an explosion. Keep an eye on the vapor cloud, for it will not dissipate rapidly until it warms up. Use large quantities of water on fires you wish to extinguish, on exposed cryogenic cylinders and to dilute spills. If at all possible, shut off the flow of liquid oxygen.

Liquid oxygen is a pretty pale blue. It is odorless, tasteless, and its only health hazard lies in its extreme cold. Table 7–2 lists some of its properties. It is the most widely used low-temperature gas. Half the total production is used to manufacture steels by supplying oxygen furnaces or by feeding the oxygen lance in open hearth operations. Another quarter produces various compounds for the chemical industry. Sizable amounts are used in certain explosives where organic materials are mixed with it just before detonation. Hospitals use 800-gallon (3-cubic-meter) LOX tanks, with high-pressure cylinders in reserve, in their oxygen systems. Piping carries it throughout the hospital. Only 15 percent of liquid oxygen production is used to oxidize various rocket propellants, but this is how LOX became famous.

Cryogenic Flammables

Two of the gases listed in Table 7–2 are flammable, adding the chance of explosive ignition to the other troublesome properties of cryogenic fluids.

The first of these, liquid hydrogen, is transparent, odorless, and the lightest known liquid, only 1/14th as heavy as water. This is one reason why it is produced in large volumes for use as a rocket propellant. By cutting the weight of fuel, it allows rockets to carry more payload. Like other cryogenic fluids, liquid hydrogen is in increasing industrial demand because this is such a convenient way to store the gas. Liquid hydrogen boils at −423°F. (−253°C.). Only liquid helium is colder. This temperature is so low it cannot only liquefy but *solidify* oxygen from the air, and solid oxygen can lodge in places where subsequent expansion can cause trouble. If it solidifies inside the liquid hydrogen itself, the combination becomes a high explosive.

Tests conducted by the Arthur Little Company for the United States Air Force produced some interesting conclusions about liquid hydrogen. It is not as dangerous to store and handle as the same amount of LP gas. Principally, this is because LP gas gathers in a cloud while hydrogen rapidly rises and dissipates. But, remember that liquid hydrogen, at its boiling point, produces an intensely cold gas which, for a time, is *heavier* than air. As the gas warms, it begins to rise at five feet per second. Quite a large vertical area can be within the flammable range of 4 to 75 percent.

A small leak, however, is likely to dissipate as fast as it is produced. You cannot pour liquid hydrogen from a small beaker fast enough for it to hit the ground. It flashes into gas almost instantaneously. Thirty-two gallons (121 liters) of liquid hydrogen will evaporate in 30 seconds. In a fire test with 5000 gallons (19 cubic meters) of spilled liquid hydrogen, flames reached 150 feet (46 meters) into the air. Storage tanks ten times larger than this are increasingly common. Considering its lightness and rate of dissipation, liquid hydrogen should be stored and handled outside whenever possible, away from any overhead enclosures.

In a bulletin ("Highway Emergency Training on the Transportation of Extra-Hazardous Cargo") the AIA had these comments about cryogenic fluids.

> Transportation of flammable cryogenic fluids, such as hydrogen and methane, presents the danger of the release of highly flammable gases in the event of an accident. Liquid oxygen, although not a flammable material, requires extra precautionary measures because it so vigorously supports combustion.
>
> Cryogenic fluids, with the exception of hydrogen, transported at pressures less than 25 psig are not subject to DOT regulations. If, however, they are shipped at greater pressures (100-200 psig), then they are classified as compressed gases and come under the regulations. Liquefied hydrogen requires special permits when shipped in bulk lots, as specified in DOT Regulation 173.316. These shipments are made in tank trucks, carrying 12,000 to 12,500 gallons, or in 30,000-gallon railroad tank cars [45 to 47 cubic meters, or in 112-cubic-meter railroad tank cars]. The vehicles should be equipped with spark arrestors and static grounding cables should be used during transfer operations. Safety valves should be set for 17 psig. If the tank pressure should approach this setting, the vehicle is to be directed off the highway, preferably to an isolated area and the excess pressure relieved through a manual blow-down valve.
>
> Should any of these materials become involved in a motor vehicle accident during shipping, it is important to keep all unauthorized personnel a safe distance away and, if possible, shut off the liquid flow. In the event of exposure to a fire, water should be played on the tank to prevent pressure build up. Spills should be allowed to vaporize with all ignition sources kept away from the area for at least one hour after the frost has disappeared from the ground. All vehicles transporting cryogenic liquids should be approached with caution if involved in an accident, regardless of whether they carry a nonflammable label or not, especially if the low-temperature liquid is leaking.

The use of liquefied natural gas (LNG) may soon exceed that of liquid oxygen. Natural gas, liquefied during periods when pipeline transmission is not at capacity, can be stored for vaporization the following winter as a supplemental supply. LNG has impurities which give it a boiling point slightly different than methane, which boils at −259°F. (−162°C.). Several LNG storage tanks are being built which will contain 175 thousand barrels (28,000 cubic meters) of liquid, the equivalent of 600 *million cubic feet* (17 million cubic meters) of gas. They will be the largest cryogenic tanks in the world. The potential fire hazards of these tanks can be judged from the following report of the explosion of a tank in Cleveland, Ohio, in 1944. Because of wartime restrictions, this disaster never received a great deal of publicity. The following description is taken from AIA Bulletin 211.

> The East Ohio Gas Company of Cleveland also stored their natural gas by liquefying it in four large tanks having a total capacity of 176 thousand cubic feet [5000 cubic meters] of liquid. One of the tanks developed a crack in a plate near the bottom as it was being filled for the first time with cryogenic fluid in May, 1943. The crack was welded and protected by a welded patch. This is the tank which gave way about a year later.
>
> It is felt a leak developed somewhere in the shell of the inner container. The escaping liquid penetrated the rock wool insulating materials which filled the space between the inner and outer shells, finally coming in contact with the outer shell. The differences in temperature brought about immediate vaporization and increased pressure. No provision was made for venting this pressure. Modern containers contain a pressure relief device to vent the space between the inner and outer shells when it is necessary.
>
> On the afternoon of October 20, 1944, several persons heard a rumbling noise in the vicinity of the tanks. At about the same time, white vapor was noted floating about 10 or 12 feet [3 or 4 meters] from the ground in the same area. Very shortly thereafter, an explosion occurred which scattered burning liquid gas and burning vapors over a large area. A second tank was ruptured creating more explosions and spreading the fire. Liquid gas found its way into sewers. The resulting explosions blew manhole covers into the air and broke water mains. It took 10 hours to bring the fires under control.
>
> Seventy-nine frame dwellings were destroyed, many others damaged; 200 parked automobiles were completely destroyed, and 11 factories seriously damaged, The intense heat—estimated at 2000 degrees F. [1093°C.]—and explosions damaged property within a diameter of 14 blocks. The property loss was $6 million.
>
> Many people were trapped in their homes surrounded by burning liquefied methane which spread so rapidly that many of them had no time to escape. One-hundred thirty-six people were killed.

QUESTIONS ON CHAPTER 7

1. What are three hazards associated with cryogenic gases?
2. What precautions should be taken before entering a truck using liquid nitrogen refrigeration?

3. What are the health hazards of nonflammable cryogenic gases?
4. What is the most common flammable liquid gas?
5. How does liquid oxygen (LOX) increase the burning rate of materials?

BIBLIOGRAPHY

Union Carbide Corp. (Linde Division):

"Precautions and Safe Practices for Handling Liquid Nitrogen."

Chicago Bridge and Iron Company:

"Cryogenic Storage Vessels."

NFPA Quarterly:

"Handling Cryogenic Fluids," by R. M. Neary.

American Insurance Association:

Special Interest Bulletins:

"The Warren Petroleum Corporation Propane Fire and Explosion" (July 7, 1951).

No. 211, March 15, 1945, "East Ohio Gas Company Explosion and Conflagration."

No. 252, revised 1963, "Comparative Hazards of Common Refrigerants."

DEMONSTRATIONS

1. If possible, get some liquid nitrogen. It is readily available, cheap, and the safest cryogenic fluid. DO NOT use liquid oxygen for any demonstration. Formation of the vapor cloud from liquid nitrogen is immediately apparent.
2. Due to the intense cold of cryogenic fluids hollow rubber balls held in the liquid will pop like light bulbs when thrown against a wall or on the floor. A banana held in the liquid for 30 seconds will drive a nail into *soft* wood. A small amount poured into a beaker will boil off rapidly, leaving a thick frosting on the outside of the glass. By pouring only a few drops into a balloon (use a funnel) you can show the liquid-vapor ratio as the balloon expands and bursts. Practice a little before class; *be sure to wear goggles and gloves.*

8 Flammable Solids

Invariably, it is vapors or gases which burn, not liquids or solids. This is why a proper concentration of flammable gas is always ready, willing, and able to explode, and why flash points are so important. You will remember that a flash point is the minimum temperature at which a liquid gives off vapor sufficient to form an ignitable mixture with the air near the surface. In other words, it is the temperature at which flammable liquids generate enough vapors to support combustion.

Solids also produce vapors. Some even pass directly from a solid to a gas. This is called SUBLIMATION. If a solid melts at its ignition temperature, any liquid state that may occur is so brief that it is not worth mentioning. Melting points and boiling points become one and the same.

Ordinary flammable solids such as wood or paper have no liquid states and no flash points. They produce flammable vapors only at their ignition temperatures. These vapors are consumed by fire as rapidly as they are formed. Indeed, without their formation and consumption, no fire is possible. We stop the flow—"closing the valves," so to speak—by cooling such a solid below its ignition temperature. This not only extinguishes the fire, but automatically halts vapor production. Just as the flammability of a liquid is determined by its flash point, and extinguishing methods must be related to this factor, the ignition temperature of ordinary solid materials is the key to their fire control. Most woods, papers, cardboards, and fabrics ignite in the neighborhood of 400 to 800°F. (204 to 427°C.). Since water is hundreds of degrees below these temperatures, the use of fog, a form of water which absorbs heat very efficiently, will rapidly gain control of fires in most solids, if applied in sufficient amounts to the right places. Sounds easy, but . . .

The exact ignition temperature of an ordinary solid depends on several variables, one of which is its moisture content. We have all seen this during grass fire seasons. A patch of damp grass, impossible to ignite with a match, burns readily when the whole field goes up. In spite of the dampness, this greater amount of heat vaporizes enough moisture to allow the temperature of the dried grass to be raised to ignition levels. But, in a section of still wetter grass, the fire goes out. It expends too much of its energy evaporating the additional moisture. Ignition temperature is never reached. During a fire, water can be employed to achieve the same result. By wetting and cooling exposed solids, we hold them below their ignition temperatures.

The flammability of ordinary solids also depends upon their shape. Solids burn on their surfaces because these are the only parts that obtain sufficient oxygen. Finely divided solids, having more surfaces available, burn more rapidly and are easier to ignite. A good example of the difference is the way we start an ordinary campfire. Shavings and kindling come first. Shavings can ignite at temperatures 300 degrees F. (167°C.) lower than those required for a large piece, such as a log, making it possible to get the fire started with a small piece of paper or even one match. A log does not immediately ignite because it conducts heat away from the source so rapidly that ignition temperatures are not reached until the entire large piece is heated. This principle is particularly true for solids that conduct heat readily, such as the flammable metals.

If subjected to temperatures of 350°F. (177°C.) for 45 minutes, some woods will ignite. They will also ignite in less than two minutes if the temperature is raised to 600°F. (316°C.). But, there is a lower time limitation. No matter how hot, a heat source must be in contact with a flammable solid long enough to raise it to its ignition temperature or no ignition will occur. This is why people can drop cigarettes into their laps without causing an immediate disaster. Although the cigarette tip is well above the ignition temperature of most fabrics, it is not allowed to remain in

TABLE 8–1. FIRE PROPERTIES OF CAMPHOR AND NAPHTHALENE

	MELTING POINT	BOILING POINT	FLASH POINT	IGNITION TEMP.	LOWER FLAMMABLE LIMITS (% BY VOL. IN AIR)	UPPER FLAMMABLE LIMITS (% BY VOL. IN AIR)	SPECIFIC GRAVITY (WATER=1.0)	VAPOR DENSITY (AIR=1.0)	WATER SOLUBLE
Camphor ($C_{10}H_{16}O$)	354° F. 179°C.	408° F. 209°C.	150° F. 66°C.	871° F. 466°C.	0.6	3.5	1.0	5.2	No
Naphthalene ($C_{10}H_8$)	176° F. 80°C.	424° F. 218°C.	174° F. 79°C.	979° F. 526°C.	0.9	5.9	1.1	4.4	No

```
        H     H
        |     |
  H     C     C     H
   \  //  \  /  \\  /
    C      C      C
    |      ||     |
    C      C      C
   /  \\  /  \  //  \
  H     C     C     H
        |     |
        H     H
```

FIGURE 8–1. Naphthalene ($C_{10}H_8$)

contact long enough. When this happens, people move fast. But, when they fall asleep smoking that last cigarette of the day, there is plenty of time.

Wood and paper produce flammable vapors only at their ignition temperatures. The melting point, the boiling point, the flash point, and the ignition temperature of the paper these words are printed on, are one and the same. All are reached simultaneously. But look at the variation in a solid like naphthalene. (See Table 8–1.)

Naphthalene and Camphor

NAPHTHALENE, either a white crystalline solid or flake which smells like mothballs, was used for this purpose for many years.* Moths hate the smell of naphthalene. But so do human beings, so it is being replaced by other smelly materials which moths hate but human beings do not.Even at normal temperatures naphthalene gives off fumes, as those who have ever opened an old trunk will testify. As temperatures rise above normal, so does the amount of vapor in the air. At 174°F. (79°C.) the percentage is 0.9 percent, naphthalene's lower flammable limit. If an ignition source is present, the time is ripe for it to be effective. If not, naphthalene will melt at 176°F. (80°C.), two degrees higher, and the liquid will continue to manufacture vapors. At 424°F. (218°C.), liquid naphthalene will boil into a gas. This gas, if heat is maintained, will ignite at 979°F. (526°C.). This is also true for camphor. The major variation is that camphor has a flash point more than 200 degrees F. (111°C.) lower than its melting point. Camphor melts at 354°F. (179°C.) but has a flash point of 150°F. (66°C.).

Remember the ring hydrocarbons, benzene, toluene, and the xylenes? Naphthalene is the big brother of this aromatic series. It is a solid for the same reason the larger molecules of the paraffin series are solids; it packs too much molecular weight to be a liquid. The molecule, a double benzene ring, is diagrammed in Figure 8–1.

---

*Although naphthalene has such names as white tar or tar camphor, it will be known as "mothballs" for a long time.

Naphthalene is shipped in glass bottles, cans, burlap bags,* or molten in tank barges. It should be stored in a cool place away from ignition sources or oxidizing agents. Naphthalene can be found where dyes, fungicides, explosives, resins, and other materials are being manufactured.

Fire-fighting techniques include the use of CO_2, dry chemicals, or ordinary foam. Although water can be used to cool the solid or liquid below its flash point, we can expect extensive frothing and foaming when water contacts hot liquid naphthalene. It may be useful to know that the liquid is heavier than water. Naphthalene vapor is poisonous and irritating to the eyes and skin. Avoid contact with the liquid and wear a self-contained mask and skin protection.

CAMPHOR is a white or colorless crystalline substance with an odor as familiar as naphthalene. Commercial camphor is produced by the steam distillation of 50-year-old camphor laurel trees, native to Formosa, but now cultivated in Florida and California. However, most of the camphor used in the United States is synthetic and slightly different from natural camphor; fifty years is a long time to wait. Camphor is used in medicines, in cellulose-nitrate plastics, in insecticides, tooth powders, and as an embalming fluid. The vapor can be irritating. Camphor is shipped in containers ranging from one-pound (.45-kg) tins to 250-pound (114-kg) barrels or slabs. Storage regulations and fire-fighting techniques are similar to those applicable to naphthalene.

Carbon

Fires in ordinary flammable solids, in wood and paper and grass, have been and will continue to be the principal reason for the existence of organized fire departments. All the flammable materials we have examined thus far, the Class "A" solids and Class "B" liquids, the woods and hydrocarbons, the papers and alcohols, the grasses and gases, the carbohydrates and ethers, have something in common. They burn, and they contain hydrogen or carbon. It has been estimated that over a million organic compounds (compounds containing carbon) have been identified. Pure elemental carbon takes many forms, some of them extremely useful to American industry. Let us see what they are.

*Some materials shipped in solid form can melt at normal or slightly raised temperatures. The problem of containment then arises. A material you are expecting to stay put begins to trickle around. The difference between a liquid and a solid is difficult to pinpoint anyway. What, for example, would you call a heavy grease or jelly? They will flow in their own good time. These types of materials, often called semi-solids, could just as logically be called stiff liquids.

Coal

The use of COAL as a fuel and raw material is familiar to everyone. Coal is amorphous (shapeless) elemental carbon.* Most coal beds began about 250 million years ago with an ancient forest of trees and ferns. Through millions of years they died and were replaced by other vegetation which died in its turn. A mattress of dead vegetation, hundreds of feet thick, built up. The slow geological evolution of the earth covered this organic matter with blankets of earth and rock. Compression started, heat increased. Vegetation is principally cellulose, which is about half carbon with varying amounts of hydrogen, oxygen, and other trace elements. Gases were slowly forced from the bed, raising the percentage of remaining carbon until lignite, or brown coal, was formed. Lignite still has a high moisture content, and parts of the original plants can sometimes be seen, but its carbon concentration is up to 60 percent.

As pressure and heat were maintained or even increased by further geological changes, bituminous or soft coal was created. There are many grades of bituminous coal, some of which have a carbon content of 86 percent. Many soft coals retain a woody structure. Finally, in beds where compressing pressures were especially intense, hard coals with a carbon percentage as high as 98 percent were formed. The amount of vegetation compressed to form a workable bed of hard anthracite coal, 30 feet (10 meters) thick, is simply stupendous. A one-foot (31-centimeter) layer of coal requires 10,000 years to develop. (The hardness of coal is not necessarily determined by age. Some very ancient beds are still lignite because not enough pressure has been applied over the course of time. PEAT is carbonized, dead vegetation compressed only by its own weight and water. A peat bog will never make the anthracite grade unless some geological event covers it with a mountain.

Charcoal

If you decide not to wait some 250 million years for a black solid to burn in your fireplace, there is a way to speed up the process: Make a pile of wood, cover it with earth, and set it on fire. In the absence of air, combustion will not be complete. The more volatile gases will be driven off, leaving you with a bed of almost pure carbon, called CHARCOAL. Charcoal is made

---

*Before you protest that coal has a very definite shape and produce a chunk to prove it, "amorphous" describes the internal lineup of atoms. Most solids are crystalline, with their atoms forming definite patterns or lattices, or are amorphous, with atoms hooked up every which way. Although carbon atoms will also crystallize, coal is amorphous (shapeless) carbon.

commercially from several carbonaceous (carbon-containing) materials. Animal bones are heated to obtain BONE BLACK, for example. The charcoal we use on weekends to burn our steaks is made from various types of wood by modern processes that heat wood to make charcoal, while preserving the valuable gases which boil off. (This same process, using bituminous coal instead of wood, creates coke, a form of carbon which burns with little smoke and such intense heat it will smelt iron.)

Charcoal has many uses. It will decolorize, deodorize, and filter materials. It is a component of black powder and other explosives. When you heat charcoal to high temperatures, a porous structure with a high capacity for absorbing vapors is created: activated charcoal. Filter-type gas masks contain activated charcoal. The porous structure allows tiny air molecules to pass through but traps some of the larger poison molecules.

Carbon Blacks

When hydrocarbons burn, they create quantities of black smoke which are mainly unburned carbon particles. Whenever anything is going to waste, somebody will try to figure out a profitable use for it. CARBON BLACK is a powder created from the smoke particles formed when natural gas or liquid hydrocarbons undergo incomplete combustion or thermal decomposition. There are several types.

1. CHANNEL BLACK is made by allowing natural gas flames to strike iron plates from which the soot is eventually scraped.
2. FURNACE BLACK is produced by the incomplete combustion of natural gas or aromatic hydrocarbons in closed furnaces.
3. A similar process, using heavy grade oils, creates LAMPBLACK.
4. ACETYLENE BLACK results from the incomplete combustion or thermal decomposition of our old friend, the triple threat acetylene.
5. Other hydrocarbons are decomposed to form THERMAL BLACK.

Carbon black strengthens natural or synthetic rubbers and will greatly increase the wear of automobile tires. A typical tire contains five pounds of carbon black and ten pounds of rubber. In fact, anything black probably includes one of these powders, for they pigment inks, dyes, paints, shoe polishes, carbon paper, crayons, typewriter ribbons, and phonograph records. Lampblack is used in insulation systems, explosives, and matches. Acetylene black absorbs liquids more effectively than other carbon blacks and has a higher electrical conductivity. It will convey these qualities into plastics and rubber, or into dry cell batteries. What with one use or another, over two billion pounds of carbon black are used annually in the United States. That's a lot of smoke.

Coal, charcoal, and the carbon blacks are black and amorphous. But carbon can also fashion crystals. If the atoms arrange themselves into a six-sided figure, called a hexagon, the result is a slippery, shiny, black solid called GRAPHITE. There are large natural deposits of graphite in Sri Lanka and Mexico, and it can be produced synthetically. Powdered graphite is a familiar lubricant. Mixed with clay, graphite is used in "lead" pencils. Pure graphite rods control the reaction in some atomic piles. Graphite powder can be used as an extinguishing agent for certain metal fires because its ignition temperature is quite high: about 1300°F. (704°C.).

Carbon has another crystalline shape. If subjected to tremendous heat and pressure, carbon atoms will realign themselves into an eight-sided crystal, called an octahedron. This form of carbon is the hardest naturally occurring substance known: a DIAMOND.* Although diamonds can be black (a tough, compact, black diamond called carbonado is often used for industrial purposes), they also come in crystals of many colors. Besides gemstones, there are diamond glass cutters, drills, wire dies, and abrasives. A diamond is pure carbon. It will burn. A conflagration in a diamond collection would undoubtedly be the smallest large loss fire in a department's history.

Hazards of Carbon

Now that we have sorted out the industrially important types of pure carbon, let us see how they concern us. As always, the hazards of a material arise out of its properties. Carbon atoms "hook" together very efficiently. (See valence, Chapter 1.) Inside a lump of carbon, most of the atoms are firmly attached to four other carbon atoms. The difficulty comes at the edge of the lump where all these "hooks" can be visualized as sticking out, waiting to latch on to something, generally a passing oxygen molecule. When they combine, carbon is oxidized and spontaneously heats. This form of heating is most prevalent in amorphous carbon. (See Figure 8–2, page 166.) When carbon atoms are crystallized, fewer possibilities exist in their ordered linked patterns for unsatisfied valences.

With the possible exception of anthracite, all grades of coal are subject to spontaneous heating and ignition. Anything which creates fresh reactive surfaces can cause trouble. As coal is mined or shipped or stored, it is broken apart accidentally, or on purpose, into smaller and more usable pieces. New surfaces are free to oxidize, and more surface is available. Since spontaneous heating is a cumulative process, coal is most dangerous 90 to 120 days after mining. Foreign matter can contribute to this danger in two ways: First, it can produce gases that are picked up by the carbon along with oxygen. This combination oxidizes together. Such a mixture may have a lower ignition temperature or may heat more rapidly than pure carbon.

---

*When an element has more than one form, the forms are called ALLOTROPES Diamond and graphite are allotropic forms of carbon.

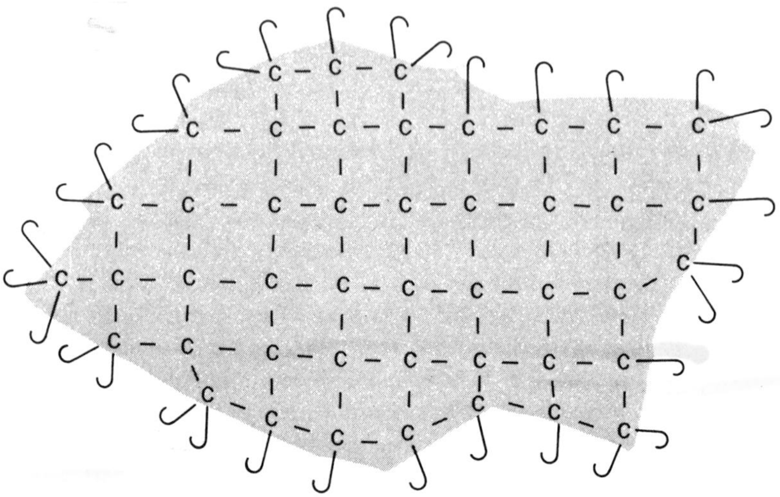

FIGURE 8–2. Schematic—Amorphous Carbon

Second, heating, an inevitable by-product of oxidation, may cause the ignition of such materials as grease, oil, paper, wood, leaves, or other debris with low ignition temperatures. For these reasons, coal should be stored in a clean location.

As we have learned from our overworked "pile of oily rags" example, spontaneous heating is a cyclic process. More heat means faster oxidation, which, in turn, creates still more heat. Coal should not be stored near heat sources which can give this process a boost. Also, just as a tightly packed pile of rags is less likely to heat spontaneously, tightly piled coal is less likely to heat. What is more important to us is that alternate wetting and drying of coal aids spontaneous ignition. Here the analogy with our rags breaks down. Coal is a solid which resists penetration by water. Only its face is washed, cleaning off the surfaces and getting them ready for action again. This is also true for charcoal, another potential spontaneous igniter.

Charcoal heats most easily when it is freshly made or ground and exposed to air, or when it is moistened and later dries. Wet charcoal should be stored separately from dry, and inspected frequently. (When wood is exposed to low-temperature heat sources for a long time, charcoal sometimes builds up on its surface. Spontaneous heating of this porous charcoal is suspected as the cause of some mysterious ignitions.)

Carbon blacks are most hazardous immediately following manufacture. This is not only because of oxidation, but because bags sometimes contain hot carbon particles. For this reason, carbon black is stored in an observation warehouse before final disposition. After cooling and airing, it does not spontaneously heat. Lampblack, considered the most susceptible to heating, often ignites spontaneously when freshly bagged. Any form of finely divided pure carbon, whether it be pulverized charcoal, coal dust, or one of the carbon blacks, is a possible dust explosion hazard. Once again, lampblack is particularly hazardous. Coal is sometimes sprayed with a high-flash-point mineral oil that not only reduces dustiness but also protects coal surfaces against oxidation.

Surprisingly, with all the problems it creates, pure carbon is considered a relatively inert element. It ignites in air at temperatures between 600 and 1400°F. (316 and 760°C.). When it combines with oxygen, the carbon atom itself does not vaporize. It merely transfers its allegiance to a carbon monoxide or carbon dioxide molecule. To turn carbon into a gas requires temperatures above those produced by normal burning. Carbon starts to sublime (turn directly from a solid to a vapor without bothering about the liquid state) at 6400°F. (3538°C.). At lower temperatures, solid carbon combines directly with atmospheric oxygen to cause a glowing combustion, one of the rare exceptions to the rule that only gases or vapors will burn. With no flame to guide us, the burning process of carbon can be very subtle. Overhaul procedures after a carbon fire must be extremely thorough.

Fighting fires in coal or charcoal can be difficult and dangerous. Water should not be used recklessly. If a small amount of coal or charcoal is involved, it may be possible to separate it and thoroughly wet it down. Otherwise, you may have no other choice but to use quantities of water to reduce the entire pile below its ignition temperature. Remember that pure carbon burns hot. There will be a possibility of a steam explosion within a heated pile. Recognize that all the wet material will have an increased ignition hazard when it dries. Your advice to the storer or shipper must include consideration of this fact.

When burning, pure carbon produces quantities of carbon monoxide and carbon dioxide. If moisture is present, steam and hydrogen gas can also form. Hydrogen and carbon monoxide are flammable; carbon monoxide is poisonous; carbon dioxide is an asphyxiant. *Pure carbon fires, therefore, are potentially poisonous or explosive, especially if they are indoors* where concentrations of gas can build up. Protect fire fighters with self-contained gas masks.

Phosphorus

As we examine the fire properties of hazardous materials, we come inevitably to one with an extremely low ignition temperature, PHOSPHORUS. Too reactive to be found free in nature, this element was first discovered by an ambitious German alchemist when he analyzed the solid residue in a sample of urine. Phosphorus has three allotropic forms. Although each is pure phosphorus—nothing added or taken away—they have distinctive appearances and properties. The industrially important allotropes are white (or yellow) phosphorus and red phosphorus (Table 8–2, page 170). (Black phosphorus, which looks like graphite, is quite rare.)

White Phosphorus

WHITE PHOSPHORUS, when exposed to air in a solid state, ignites at 86°F. (30°C.); when finely divided, it ignites at room temperature. At normal temperatures, solid white phosphorus combines with atmospheric oxygen and glows. It is "phosphorescent."

How can industry store and handle such a material? Fortunately, phosphorus will not normally ignite when cut off from atmospheric oxygen. It can, therefore, be shipped in hermetically sealed cans, in drums, or in tank cars under either water or an inert gas. Extreme care must be used in handling these containers. Once white phosphorus is exposed to air, ignition is very likely. In storage areas, there should be a periodic check for water leakage. Phosphorus should be stored in a protected area, apart from other materials, particularly oxidizing agents. When mixed with such materials, it can explode.

White phosphorus is a colorless or white waxy solid, often found in sticks about the thickness of a man's thumb. It turns yellow on exposure to light. Solid white phosphorus causes severe flesh burns on contact. Do not let it touch the skin. Inhalation or ingestion of even small amounts (1/300 of an ounce or 850 milligrams) can be fatal. Fumes from burning phosphorus, while not considered highly toxic, are very irritating and extremely dense. Flame-proofed protective clothing and a self-contained mask are highly recommended for firemen fighting a white phosphorus fire. The key to extinguishment is the fact that phosphorus will not burn under water. It is almost twice as heavy as water and not soluble. If you drown it with large amounts of water without allowing hose streams to spread the liquid phosphorus, it will solidify below 112°F. (44°C.) and can be covered with wet sand or dirt. Be careful about the premature removal of this cover. Re-ignitions are very likely. Once you have the phosphorus extinguished, pause and consider your next move.

Industrially, white phosphorus is used in certain delayed-action rat poisons, incendiary grenades, and bombs. The use and effects of the 30-pound white phosphorus bomb during World War II have been graphically described by Martin Caidin:*

> The U.S. Strategic Bombing Survey reports officially that the psychological effect of phosphorus bombs was far greater than any

---

*Martin Caidin, *The Night Hamburg Died* (New York: Ballantine Books, 1960).

> actual damage they caused. British reports of phosphorus bombs dropped by the Germans against English cities indicate that the use of this incendiary material was severe enough in its effect to drive the fire-fighting teams crazy. Whenever it dried out, it immediately exploded into flames. You could cover phosphorus with sand and douse it with water and it would stay quiescent. But the moment the covering was gone, the flames immediately burst out again!
>
> Phosphorus sticks grimly to any surface it touches—metal, wood, concrete, clothing, or human flesh. Whatever the material, the phosphorus will cling to it for good (unless it is scraped off) and will continue to burn itself out unless it is permanently denied access to oxygen.

Caidin's descriptions of human beings caught in a shower of white phosphorus near a river is unforgettable.

> Those with the flaming chemical on their arms and legs and bodies were able to extinguish the flames instantly on entering the water. But those agonized souls who had the phosphorus on their faces and heads! Certainly the flames went out as they plunged into the waters of the Alster. The moment they came up, however, the phosphorus received its oxygen and again burst into flames. And so began the unbelievable terror of choice—death by drowning or by burning.

Red Phosphorus

RED PHOSPHORUS is much less dangerous than white phosphorus. Most important, its ignition temperature is above 392°F. (200°C.), although this temperature can be easily reached in a fire. Under some fire conditions, it is possible for red phosphorus to revert to white phosphorus. The same care must be exercised against re-ignitions. Again, the use of large amounts of water is recommended. It will produce quantities of dense fumes that should be avoided. Use self-contained masks. Red phosphorus does not have to be shipped under water, although both forms carry the same DOT red label as flammable solids.

Red phosphorus is preferred by industry as a source of pure phosphorus. It is used in the manufacture of phosphoric acid and other phosphorus compounds such as phosphate fertilizers and insecticides. In one form or another, 600 million pounds (273 million kg) of phosphorus are used every year. You probably carry red phosphorus in your pocket most of the time, proving that it does not have the immediate corrosive effect of its allotropic kin. One use of red phosphorus is in the manufacture of matches. It is mixed with glue and ground glass to provide the striker for safety matches. When a safety match is drawn across the surface of this striker, a tiny piece of the

TABLE 8–2. PROPERTIES AND LABELING OF WHITE AND RED PHOSPHORUS

	IGNITION TEMPERATURE	MELTING POINT	BOILING POINT	SPECIFIC GRAVITY	VAPOR DENSITY	WATER SOLUBILITY	DOT LABEL REQUIREMENTS
White phosphorus	86° F. 30°C.	112° F. 44°C.	549° F. 287°C.	1.8	4.3	None	(Dry) Flammable solid and poison label. Shipment by passenger carrying aircraft or railcar or cargo only aircraft is forbidden. (In water) Same as above except that it may be carried in packages not to exceed 25 pounds (11 kg) per package, on cargo only aircraft.
Red phosphorus	(over) 393° F. 200°C.	——	781° F. 416°C.	2.3	4.3	None	Flammable solid label required. Shipment on passenger carrying aircraft or railcar is forbidden. Packages not to exceed 11 pounds (5 kg) per package may be carried on cargo aircraft.

This red-and-white-striped label is required on containers when flammable solids are shipped. The flammable solid placard for use on motor transports and rail cars is similar in design and color.

FIGURE 8–3. DOT Flammable Solid Label

red phosphorus ignites by friction and fires the flammable mixture on the tip of the safety match. Ordinary strike-anywhere matches have other phosphorus compounds with a low ignition temperature built into their heads. One such compound is PHOSPHORUS SESQUISULFIDE.

Phosphorus Compounds

Phosphorus sesquisulfide is the first flammable compound we have discussed that contains neither hydrogen nor carbon. The molecule, a combination of four phosphorus atoms and three atoms of sulfur, P_4S_3, is a highly flammable mass of yellowish crystals. Its ignition temperature is the same as the boiling point of water, 212°F. (100°C.). Of course, phosphorus

This label must be applied to all packages containing Class B poisonous materials. It has a white background with black lettering and a black skull-and-crossbones symbol. The poison placard that must be shown on motor transports and rail cars is similar in design and color.

FIGURE 8–4. DOT Poison Label

sesquisulfide ignites easily by friction. A thumbnail will light a match, and so will ignition sources not otherwise considered dangerous, such as steam pipes. Sesquisulfide is not considered a severe toxic itself, but it produces sulfur dioxide when it burns. Use self-contained masks. Like pure phosphorus, sesquisulfide is heavier than water and not soluble. Fire fighting, once again, can include flooding with water because blanketing is possible. The compound is shipped in glass jars and bottles, cases and steel drums under a DOT red label and a poison label. It must be shipped separate from flammable gases or liquids, oxidizing materials, or hydrogen peroxide. Storage should be in a cool, protected place.

PHOSPHORUS PENTASULFIDE, P_2S_5, a yellow crystalline substance with an indescribable odor, is used as a lubrication oil or rubber additive, or in insecticides. It has several interesting characteristics.

1. Although it has an ignition temperature of 287°F. (142°C.) and ignites easily, it does not burn rapidly.

2. It reacts with water in several ways. It is decomposed by moist air and can heat or ignite. It reacts with water to form a poisonous and flammable gas, hydrogen sulfide.

3. It is more toxic than sesquisulfide and also forms toxic and irritating gases when it burns.

Therefore, while water can extinguish pentasulfide fires, using CO_2 is preferable. Self-contained masks should be worn. Pentasulfide is shipped in glass bottles and sealed drums which should be kept closed whenever possible. It also carries the DOT red label as a flammable solid.

A well-handled fire involving this material is reported in AIA Special Interest Bulletin No. 129.

> A plant was engaged in grinding phosphorus pentasulfide, which is readily ignited by friction. To prevent ignition, the mill was kept charged with an inert atmosphere of carbon dioxide. Presumably because of the freezing of a valve, the discharge of the carbon dioxide stopped. An explosion ignited other quantities of the material (P_2S_5) in nearby drums. The fire department, on arrival, did not use water, which was a proper procedure. The fire was extinguished by men wearing gas masks and using carbon dioxide extinguishers and dry ice, but it took four hours. The fire was not severe and little damage was done to the building.

Sulfur

Like carbon, hydrogen, and phosphorus, SULFUR is an element that burns. It is generally a yellow solid, occurring most often as a crystal or powder. If sulfur vapors are suddenly cooled, they form a yellow powder known as flowers of sulfur. Liquid sulfur freezes into solid brimstone. (Brimstone is an ancient name for sulfur.) Despite its bad reputation, pure sulfur is almost odorless. But it is part of many compounds not welcome in polite society. When it burns it forms SULFUR DIOXIDE gas, which smells as terrible as sulfur is supposed to. Sulfur combines with hydrogen to create HYDROGEN SULFIDE gas, with an aroma of rotten eggs, or with carbon to make CARBON DISULFIDE, which smells like a head of wet, year-old cabbage. Hydrogen sulfide (H_2S)

carries both the DOT red flammable gas and a poison label. Carbon disulfide carries the DOT red flammable liquid label. The MERCAPTANS, a group of sulfur-containing compounds, sometimes used for odorizing natural gas, are also found in the working end of a skunk. (See Table 8–3 for a summary of the properties of sulfur. Several other sulfur compounds are also included or briefly summarized later in this section.)

As we have seen, flammable solids are very versatile in finding ways to burn. Naphthalene and camphor produce flammable concentrations or vapor below their melting points. They are solids with a flash point. Carbon has a boiling point so high the solid combines directly with atmospheric oxygen without vaporization. White phosphorus also ignites in the solid state. Elemental sulfur finds yet another way: It melts at 234°F. (112°C.). Just like flammable liquids, molten sulfur can flow and spread fires, although only between 234°F. and 392°F. (112°C. and 200°C.) it has the peculiar property of increasing in viscosity as its temperature is raised. The color of liquid sulfur darkens, and stickiness increases up to 392°F. (200°C.). Above this temperature, the color lightens again, viscosity diminishes, and flowing starts in earnest. Molten sulfur has a flash point of 405°F. (207°C.) and, within a narrow range of 45 degrees F. (25 degrees C.), sulfur vapors can be ignited by a source. At 450°F. (232°C.), sulfur's ignition temperature, a source is no longer necessary. At 832°F. (444°C.), molten sulfur boils into a gas.

Fighting Sulfur Fires

With clouds of yellow-to-orange gases, molten dark brown pools, the choking odor of poisonous sulfur dioxide, and the blue flames of burning brimstone, a sulfur fire can be a fearful sight. Small wonder this was considered a reasonable approximation of the fires of hell, and sulfur was believed the stuff from which the eternal fires sprang. Fighting of sulfur fires should include the use of fog patterns that will cool the element below its ignition temperature and somewhat reduce concentrations of sulfur dioxide. Although there is some possibility of a steam explosion, sulfur is more than twice as heavy as water and not soluble. Water should not be trapped easily. While the careful use of water fog can be very efficient, avoid the use of the straight streams. Not only will they increase the likelihood of a steam explosion, but they can also create clouds of sulfur dust that may explode. The ignition temperature of such a cloud is 75 degrees F. (42 degrees C.) lower than the solid, beginning around 375°F. (191°C.).

The toxicity of pure sulfur is quite low. The solid material has no action upon the skin, although sulfur *compounds* have been used in various ointments for skin ailments since ancient Egyptian times. Sulfur dioxide, sulfur's combustion product, is highly toxic, however, and molten sulfur can burn the skin badly. Wear protective clothing and use a self-contained gas mask.

The annual production of sulfur is well over 15 million tons and it is an element of many uses. One is to vulcanize rubber. Before vulcanization, rubber is structurally weak and easily deformed, softened by heat, made stiff by cold, and soluble in many common liquids, including gasoline. Vulcanization modifies all of these undesirable properties. Sulfur forms a great variety of compounds: sulfuric acid, carbon disulfide, sulfur dioxide, hydrogen sulfide, and others. Pure sulfur can act as a fungicide in the lime and sulfur mixture familiar to gardeners. The main use of sulfur is in the production of the so-called "workhorse of industry," sulfuric acid, H_2SO_4. It has been said that one index of a nation's economy can be found in the amount of sulfuric acid it produces annually.

This label, with black-and-white background and white lettering, is required on all shipments of corrosive materials. The corrosive placard is identical.

FIGURE 8–5. DOT Corrosive Label

Along with saltpeter and charcoal, sulfur is one of the components of black powder. If sulfur comes in contact with a pure carbon material, such as charcoal, it can spontaneously heat and ignite. Otherwise, it has little tendency to do so. In storage, because of this property, sulfur should be kept separated from oxidizers and carbonaceous materials. Storage areas should be cool, well-ventilated, and free from accumulations of dust.

Sulfur Compounds

AMYL MERCAPTAN, $C_5H_{11}SH$, is a liquid composed of a mixture of isomers. It is colorless to light yellow, has a very disagreeable odor, is a skin and respiratory irritant, and is used as an odorant for detection of natural gas leaks. It is shipped in 1- to 55-gallon drums (4 to 207 liters) with a DOT red label: flammable liquid. Foam, carbon dioxide, or dry chemical can all be effective on fires. (See Table 8–3 for other properties.)

ANTIMONY PENTASULFIDE, Sb_2S_5, often called red antimony, is an orange-to-yellow odorless powder which is combustible and easily ignited. It yields hydrogen on contact with strong acids, particularly concentrated hydrochloric, and it yields sulfur dioxide when burning. It is shipped in cans and fiber drums and is used as a pigment, in vulcanization and coloring of rubber. Use water and wear full protective equipment.

DIMETHYL SULFATE, $(CH_3)_2SO_4$, is a colorless liquid. If it touches the skin for even short periods, it can cause intense irritation several hours later. There is no odor or immediate irritation to give a warning, so it is very dangerous. Heavy exposures can cause death. Spilling of the liquid on the skin can also cause ulcers. Complete protective equipment should be worn. It is shipped with a DOT black and white label: corrosive material. (See Table 8–3.)

DIMETHYL SULFIDE, $(CH_3)_2S$, is a colorless-to-straw-colored flammable liquid with a disagreeable odor, shipped in drums and tank cars with a DOT red label: flammable liquid. It is very toxic, with high concentrations of vapor quickly reached because of its low boiling point. Use water spray, carbon dioxide, or foam on fires. Sulfur dioxide is the combustion product. Wear protective clothing and self-contained mask. (See Table 8–3.)

HYDROGEN SULFIDE, H_2S, is a colorless flammable gas with a strong offensive odor, shipped in steel pressure cylinders with a DOT red label: flammable gas, *and* a black and white label: poison gas. Typical cylinder pressure at 70°F. (21°C.) is 260 psi (18 atmospheres). Hydrogen sulfide is very toxic, and moderate concentrations can cause immediate death. Low concentrations are irritating to the eyes and respiratory tract. Major fire objectives: stop flow of gas, keep cylinders cool, dilute concentrations of this water-soluble gas with fog patterns. Wear self-contained masks. (See Table 8–3.)

TABLE 8–3. PROPERTIES OF SULFUR AND SOME SULFUR COMPOUNDS

	FLASH POINT	IGNITION TEMP.	BOILING POINT	MELTING POINT	FLAMMABLE LIMITS (PERCENT BY VOLUME IN AIR)		VAPOR DENSITY (AIR= 1.0)	SPECIFIC GRAVITY (WATER= 1.0)	WATER SOLUBLE	DOT LABEL REQUIREMENTS
					LOWER	UPPER				
Amyl mercaptan $C_5H_{11}SH$	65° F. 18° C.	—	260° F. 127° C.	−104° F. −76° C.	—	—	3.6	0.8	No	Flammable liquid
Carbon disulfide CS_2	−22° F. 30° C.	212° F. 100° C.	115° F. 46° C.	−167° F. −111° C.	1.3	44.0	2.2	1.3	No	Flammable liquid
Dimethyl sulfate $(CH_3)_2SO_4$	182° F. 83° C.	842° F. 450° C.	370° F. 188° C.	−16° F. −27° C.	1.4	7.5	4.4	1.3	Very Slightly	Corrosive material
Dimethyl sulfide $(CH_3)_2S$	0° F. −18° C.	402° F. 206° C.	99° F. 37° C.	−145° F. −98° C.	2.2	19.7	2.1	0.8	No	Flammable liquid
Hydrogen sulfide H_2S	— —	500° F. 260° C.	−77° F. −61° C.	−176° F. −116° C.	4.3	45.0	1.2	1.08	Yes	Flammable gas and poison
Sulfur	405° F. 207° C.	450° F. 232° C.	832° F. 444° C.	234° F. 112° C.	—	—	Varies	2.0	No	None
Sulfur chloride S_2Cl_2	245° F. 118° C.	453° F. 234° C.	280° F. 138° C.	−112° F. −80° C.	—	—	4.7	1.7	Decomposes	Corrosive material

LEAD SULFOCYANATE, $Pb(SCN)_2$, is a white-to-light-yellow crystalline powder, shipped in fiber and stainless steel drums which should be kept closed. It does not burn rapidly, but decomposes in air to form carbon disulfide and sulfur dioxide. The dust is a possible explosion hazard. Use water and wear self-contained masks. It is used in priming mixtures for small arms ammunition, safety matches, and the manufacture of dyes.

SULFUR CHLORIDE, S_2Cl_2, is an amber-to-yellow oily liquid with a penetrating odor. The fuming liquid is corrosive. The vapors are irritating to the eyes, lungs, and mucous membranes. Wear complete protective equipment. Sulfur chloride is shipped in iron drums inside wooden crates with a DOT black and white label: corrosive. It decomposes on contact with water. Uses include: sulfur solvent, poison gas manufacture, hardening of soft woods. (See Table 8–3.)

SULFUR DIOXIDE, SO_2, is a colorless gas with a choking odor. It is highly irritating to the eyes and respiratory tract. Low concentrations can cause death. Its vapor density is 2.3 and its boiling point is 14°F. (−10°C.). It is shipped with a DOT green label: nonflammable gas. It is shipped in steel pressure cylinders containing 200 to 800 pounds (91 to 364 kg) of liquefied gas or in large tank cars. Typical cylinder pressure at 70°F. (21°C.) is 35 psi (2.3 atmospheres). It is used industrially as a bleaching agent, food preservative, fumigant, color restorative in aged grains, and in the manufacture of chemicals. It is the common combustion product of sulfur-containing materials. Major fire objectives: stop flow of gas, keep cylinders cool. Wear self-contained masks.

Carbon Disulfide

If a mad scientist decided to create a super flammable liquid, what properties would he give the monster?

1. *A flash point and a boiling point so low flammable vapors will generate in abundance at almost any temperature.* The combination of a flash point of −22°F. (−30°C.) and a boiling point of 115°F. (46°C.) will mean vapors are produced early and often. If the material is heated even slightly, vapor pressure will be considerable.

2. *An ignition temperature so low almost any source of heat will be sufficient to explode its plentiful vapors.* The boiling point of water is about right. Although another compound we have met, phosphorus sesquisulfide, has the same ignition temperature, it is a solid and tends to stay put. As we have seen, however, once free, vapors of a liquid will travel and seek out an ignition source. At 212°F. (100°C.), an ordinary light bulb, a steam pipe, a metal roof heated by sunlight, or

any one of a dozen other sources of heat not ordinarily considered hazardous, can be that ignition source.

3. *An extremely wide flammable range so that percentages too rich and too lean are almost impossible.* A range of 1.3 to 44 percent ensures the explosive quality of almost every collection of vapor, whether caused by a small leak or a major spill. In addition, easily reached concentrations between 4 and 8 percent should explode with a maximum of violence.

4. *A high vapor density to make sure vapors stay together until they find an ignition source.* A density more than twice as heavy as air, 2.2, ensures the same type of cloud that makes the LP gases so dangerous. Of course, the liquid and vapor should *not* be water soluble. This would make it too easy for fire fighters to wash it out of the air with fog patterns. The liquid should be soluble in alcohol, ether, and benzene; none is recommended as a dilution agent in fire fighting.

5. Finally, add a few minor touches: Make it so toxic fire fighters must have their vision partially obscured by masks. Make it so cheap and useful industry will produce it in quantity. Make the liquid *mobile,* which means it will run all over the place because it has little viscosity. Not a bad recipe for a monster.

Put all these ingredients together and they spell CARBON DISULFIDE, a liquid with such a unique combination of properties that it is generally considered *the most dangerous of all common flammable liquids.* In a hazardous classification system where kerosene is rated at 40, a flammable liquid like ethyl alcohol at 70, gasoline at 95–100, and ethyl ether at 100, the Underwriters' Laboratories give carbon disulfide the gold medal rating of 110. Not content with that, they add a plus. Any 110+ liquid deserves a section of its own. (See Table 8–3.)

Small amounts of carbon disulfide occur naturally in coal tar or crude petroleum, but it is prepared on an industrial scale either by heating pure carbon (charcoal, coke, or coal) in a furnace, along with sulfur vapors, or by the interaction of sulfur and methane vapors. Since both end products, carbon disulfide and hydrogen sulfide, are useful, this process has become increasingly popular since the 1950s.

$$CH_4 + 2S_2 \longrightarrow CS_2 + 2H_2S$$

Methane **Sulfur** **Carbon disulfide** **Hydrogen sulfide**

Unfortunately, carbon disulfide has a range of properties which makes it valuable. It will dissolve phosphorus, sulfur, selenium, bromine, iodine, fats, resins, and rubbers. Toxic enough to be an insecticide or rodenticide, it is most effective as a fumigant at concentrations slightly above its lower flammable limit. The *major use* of carbon disulfide is in the manufacture of rayon and cellophane, although large quantities are employed to cold-vulcanize rubber, to manufacture carbon tetrachloride, and to make explosives.

Carbon disulfide is a colorless-to-pale-yellow liquid with a strong disagreeable odor which has been poetically described as decayed cauliflower, moldy broccoli, or rotten cabbage. The purest distillates are reported to have a sweet and pleasing aroma. High concentrations of vapor, whether sweet or foul, are narcotic and cause death by respiratory paralysis. One part in 250 in air (4000 parts per million) is fatal in one to two hours. Lower concentrations can cause symptoms of intoxication, muscular weakness, sensory impairment, and unconsciousness. *It is poisonous by oral intake, inhalation, or prolonged contact with the skin.* Wear protective equipment and a self-contained mask.

Extinguishing a Carbon Disulfide Fire

How can fire from such a monster be extinguished? We know that carbon disulfide is not soluble in water, but it has two other properties that let us formulate a sensible plan of attack. First, it burns with a very low heat of combustion, less than half as hot as gasoline; second, the specific gravity of carbon disulfide is 1.3 times heavier than water. Not much, but these properties do offer some help.

Let us examine how ordinary fire-fighting agents work on carbon disulfide. Foam is ineffective because CS_2 vapors will penetrate the blanket and burn above it; the vapor pressure caused by a low boiling point is too much for foam. Carbon dioxide is not too efficient for a different reason. Complete extinguishment requires an atmosphere of 55 percent CO_2, twice as much as with gasoline. Here the 44 percent upper flammable limit of CS_2 is in operation. Dry chemical is not a world-beater, either. Carbon disulfide vapors re-ignite easily because of their low ignition temperature. Dry chemical has little cooling effect on hot surfaces. In order to have any chance of success on a carbon disulfide spill fire, dry chemical must be employed in conjunction with cooling water spray.

This leaves us with our old reliable, water. Because of its specific gravity and low heat of combustion, a flaming tank of carbon disulfide is most likely to be extinguished through the use of water. A fairly close approach, from upwind of an involved container, favored by a very low heat of combustion, may be possible. If so, BLANKETING, floating a layer of water over the top of a high-specific-gravity liquid which is not soluble in water, may turn the trick. But blanketing must be done gently. An open hose-butt is the best applicator because it will deliver quantities of water without agitating the surface unduly. A thick enough layer of water over a carbon disulfide fire will extinguish it.

Special Storage Requirements

A material such as carbon disulfide must have special storage requirements. Here, indeed, is where an ounce of fire prevention is worth a ton of fire extinguishing agents.

Store carbon disulfide containers in an isolated, preferably noncombustible building in which they may be safeguarded against injury. The area should have no heat, direct sunlight, hot pipes, or electrical lighting fixtures. Any artificial light should shine through glass ports. Protect against static electricity. Spray drums with water during hot weather to reduce vapor pressures. Provide floor level ventilation for indoor storage. (Construct large tanks to allow for underwater storage or over concrete basins large enough to hold all the contents of the tank in addition to the water.) Keep large containers blanketed with water or inert gas at all times. Use wooden measuring sticks to check tanks, not spark-producing metals. Never dispose of carbon disulfide by dumping on the ground or into sewers, but always by burning in a safe location. It should be transferred by pump, or displaced by an inert gas or by water, rather than by air pressure.

Finally, CS_2 is shipped in small gas or metal containers (1- to 5-pound bottles or cans) inside *ordinary wooden or cardboard boxes,* in 5- to 55-gallon drums, or in tank cars up to 7000 gallons. The DOT requires a red label: flammable liquid. Carbon disulfide cannot be shipped by express. Transportation is of more than passing interest, for carbon disulfide was the liquid that caused the Holland Tunnel fire. The following description is condensed from the May 13, 1949, report of the AIA.

> At about 8:45 A.M. on Friday, May 13, 1949, a large trailer truck, loaded with 80 fifty-five-gallon drums of carbon disulfide entered the Jersey City side of the Holland Vehicular Tunnel that runs under the Hudson River to New York City. Commerical traffic was particularly heavy that morning and moving slowly. Five trucks, including the carbon disulfide unit, entered the tunnel at about the same time. About 350 feet behind them was another group of five trucks. [See Figure 8–6, page 182.]
>
> Three minutes later, a patrolling officer saw a truck, apparently stalled in the tunnel, some 2900 feet from the New Jersey entrance. As he ran toward the truck, a loud blast occurred and he was met by two

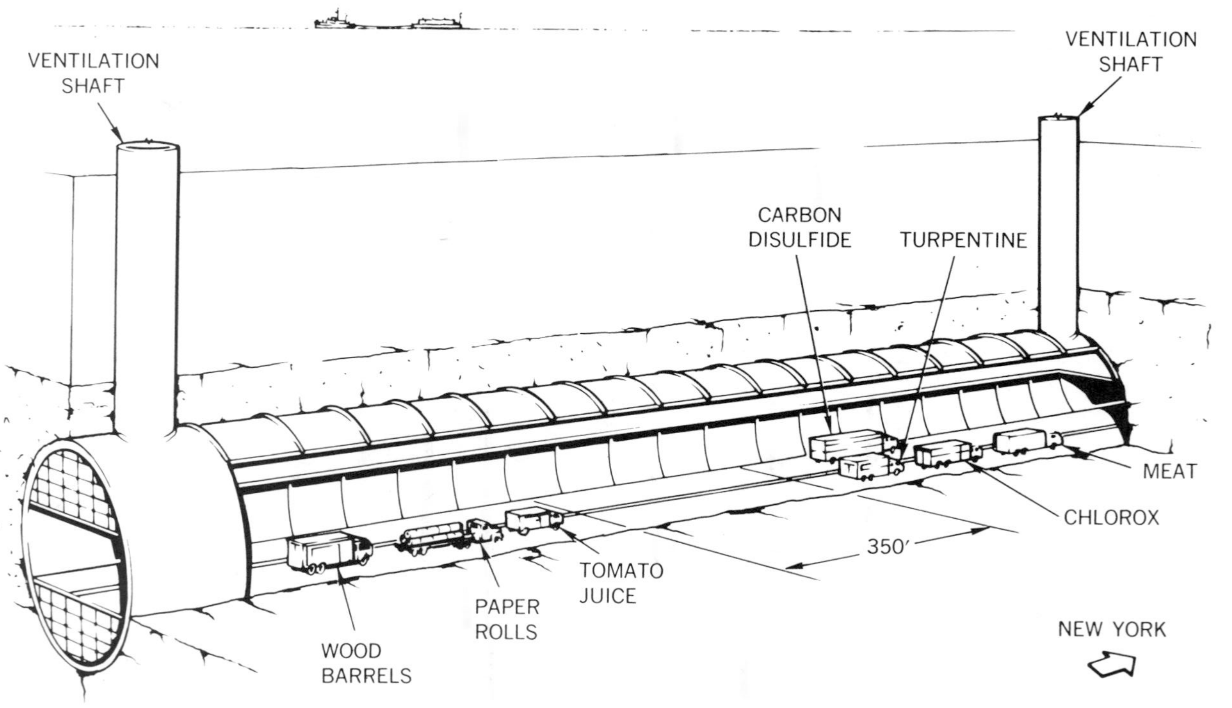

FIGURE 8–6. Holland Tunnel Fire

men. He guided them to safety. Another tunnel officer, noticing dense fumes, smoke, and no emerging traffic, gave the alarm.

Automatically, signal lights behind the emergency turned red, halting incoming traffic; signals ahead turned amber, shifting traffic to the right-hand lane and slowing it down and opening the left lane for emergency equipment. The Port Authority crew, entering the tunnel from the New York end, which was free of traffic, discovered a large ball of fire in the tube. Protected by all-service masks, they were able to connect a 1½-inch hose to a standpipe outlet about 75 feet from the first burning truck, knock down the fire in trucks 1 and 2, and advance on truck 3 when they noticed flaming liquid running down the gutters on both sides of the tunnel. The first rescue company from the N.Y. Fire Department arrived at this point and took over operation.

They found a very serious situation: a number of trucks on fire, extreme heat, bursting carbon disulfide drums, toxic fumes, and minimum visibility because of heavy smoke. In addition, broken circuits had put the tunnel lights out and damage to the tunnel ceiling threatened a possible collapse of the tube itself. Fireboats were positioned above the tunnel to watch for tell-tale bubbles, showing a leak.

So intense was the fire that it spread through 350 feet of empty tunnel to the second group of trucks, near the Jersey end. All the wall surfaces and ceiling slabs of this section were demolished; large sections of concrete were left hanging by their reinforcing bars. The roadbed was completely covered with broken concrete and tile and the trucks involved were reduced to twisted heaps of metal.

Until 125 stalled cars were removed, the Jersey City Fire Department, moving from the other end of the tunnel, had to hand-carry hose to the standpipes closest to the fires in the second group of trucks. In spite of heavy fumes, they advanced quite rapidly, putting out the nearest truck fires one at a time.

By nightfall, in spite of a re-flash of carbon disulfide fumes that was quenched by foam, the fire was under control. Cleanup required the removal of over 650 tons of debris from the tunnel. There were no fatalities but a great many firemen were hurt. The major cause of injury was smoke and gas inhalation; most of the first men from the NYFD, several of the Port Authority crew, and ten of the Jersey City men had to be treated. Wartime civil defense gas masks, pressed into use, were ineffective against the toxic gases in the tunnel.

All this damage was caused by a single truckload of carbon disulfide, showing the untold potential hazards in our normal day-to-day highway transportation of dangerous chemicals. By sheer luck, passenger cars and buses, including three buses full of children, were not in the tunnel when the fire began.

TABLE 8–4. PROPERTIES OF SELECTED FLAMMABLE SOLIDS

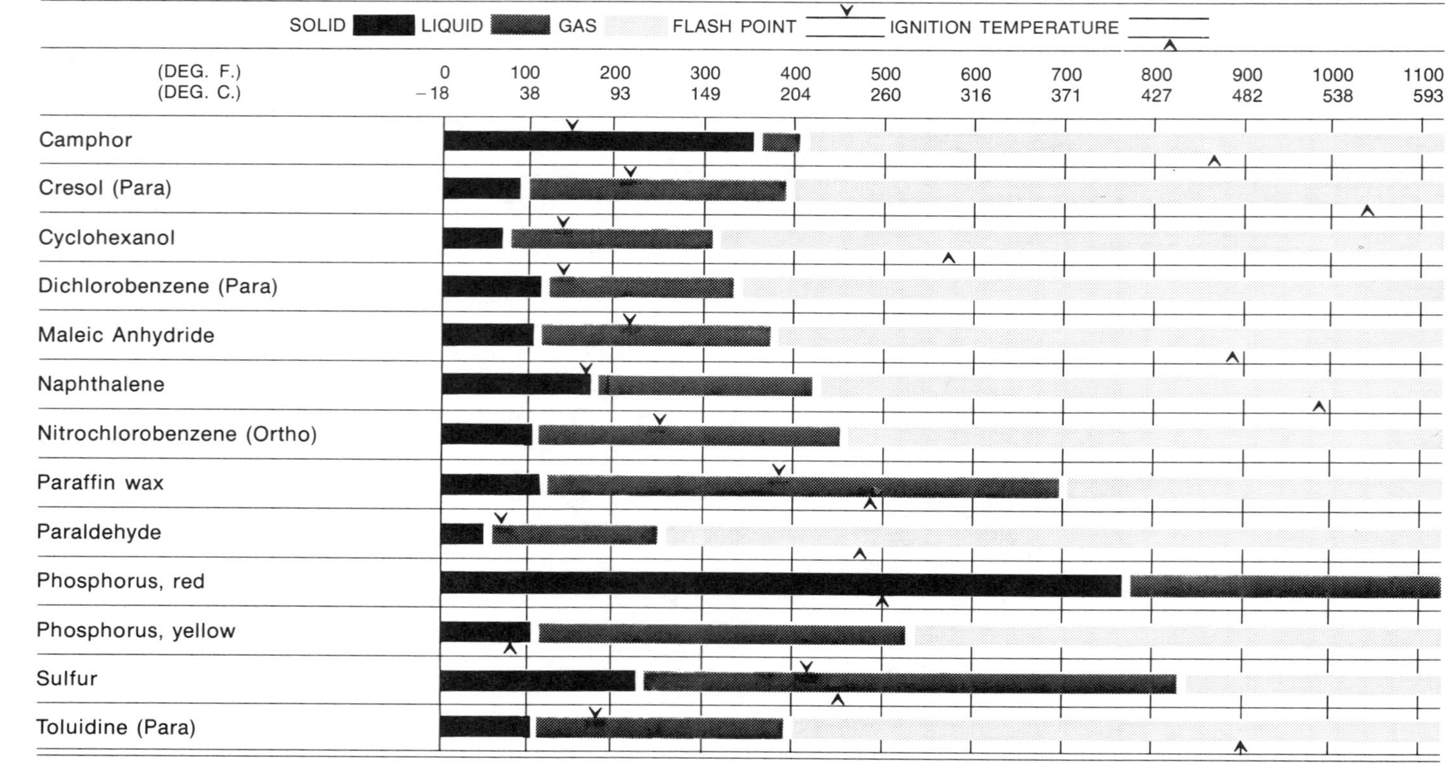

Summary

Solids find many ways to burn.

1. An ordinary flammable solid produces vapors at its ignition temperature that are consumed as they are produced.

2. A solid like camphor or naphthalene passes directly from a flammable solid to a gas. Since the ignition temperature is still higher, these vapors have time to build up to their flash point.

3. Solids like white phosphorus and carbon can ignite in their solid state. When, and how they do so, varies.

4. Like sulfur, a flammable solid can melt. The liquid then produces vapors and reaches a flash point. Flammable solids can have flash points to go along with their ignition temperatures. Ignition can occur before or after the solid has melted into a liquid or boiled into a gas. The flash point can come either in the solid or liquid state. (See Table 8–4 for examples of this versatility.)

QUESTIONS ON CHAPTER 8

1. What does *sublime* mean?
2. Name two examples of solid materials which have a flash point.
3. What is an *allotrope*?
4. Why is the use of water not advised on a fire involving a pure carbon material?
5. What are the ignition temperatures of white and red phosphorus?
6. How is white phosphorus shipped and stored?
7. Why should straight streams be avoided when fighting sulfur fires?
8. Discuss the toxicity hazards of dimethyl sulfate.
9. What effect will fog patterns have on concentrations of hydrogen sulfide gas?
10. Discuss the fire fighting of carbon disulfide.
11. When does a Class "A" solid produce flammable vapors?
12. Name two factors which influence the flammability of ordinary solids.
13. Name three types of carbon black.
14. What is the recommended extinguishing agent for a phosphorus fire?
15. Of the two phosphorus compounds discussed in this chapter, which has the lowest ignition temperature? Which is most reactive with water?

BIBLIOGRAPHY

American Insurance Association:

Special Interest Bulletins:

No. 2, revised October 30, 1950, ''Wood—Ignition Temperature and Related Matters.''

No. 30, April 15, 1953, ''Preventing and Extinguishing Fires in Soft Coal.''

No. 129, September 15, 1941, ''Delays in Fire Prevention May Be Costly.''

Report: ''The Holland Tunnel Chemical Fire,'' May 13, 1949.

Handbook of Industrial Loss Prevention, Factory Mutual Engineering Division:

''Carbon Disulfide,'' Chapter 52, 1967.

Manufacturer's Chemists Association:

Chemical Safety Data Sheet SD-74, ''Sulfur.''

VISUAL AIDS

Three movies which demonstrate the industrial uses of solid materials are available from many college film libraries: ''Carbon and Its Compounds,'' 10 minutes, color; ''The World of Phosphorus,'' 27 minutes, color; ''Sulfur and Its Compounds,'' 10 minutes, color.

DEMONSTRATIONS

1. The blue flame of burning sulfur and the odor of sulfur dioxide can be demonstrated if you take your class outside, or inside, if there are facilities for immediately exhausting toxic gases. (Ask the chemistry department.)
2. A mixture of 60 percent carbon disulfide and 40 percent carbon tetrachloride can be ignited and held in the hand, or poured on a rag which will be undamaged by the flame. This demonstrates the low heat of combustion of CS_2.
3. White phosphorus can be ignited by a burned-out match. Place the phosphorus in a Pyrex or metal dish, light the match, blow it out, and immediately place it against the phosphorus, which will quickly ignite. This demonstration should be conducted out of doors with consideration as to where the vapors of the burning phosphorus will travel. Make sure you wear gloves and use forceps when handling white phosphorus. Keep class upwind.
4. If you wish to demonstrate the ignition of white phosphorus at room temperature, dissolve the element in five times its volume of carbon disulfide. This will create a solution of finely divided phosphorus which is flammable at room temperatures. It can be poured on a surface out of doors and will quickly catch fire spontaneously as the carbon disulfide evaporates.

9 Combustible Metals

Previously we discussed the properties of eight gaseous elements: highly flammable hydrogen; oxygen, the combustion supporter; and some more or less inert gases—nitrogen, helium, argon, neon, krypton, and xenon. A large part of the previous chapter emphasized the differences between three elemental flammable solids: carbon, sulfur, and phosphorus. All these elements have one thing in common: None is a metal. This is more unusual than it may seem, because most of the other elements are metals. Of the 106 or so elements, only 22 are nonmetals. Half of this total are mentioned above; five more nonmetals will be met in a section on halogens: fluorine, chlorine, bromine, iodine, and astatine. (Astatine and radon, another inert gas, are both radioactive.) These 17 elements make up the major part of the nonmetallic group. Several others, including boron, selenium, arsenic, silicon, and tellurium, have properties which place them somewhere in between metals and nonmetals. As a result, they are often called *metalloids,* or semimetals.

This leaves us with the properties and hazards of 81 metals to learn. Although many are so rare as to rate only a brief mention, we have a lot of ground to cover.

First, let us define a metal and learn how it differs from a nonmetal. There are a series of true statements we can make about metallic elements in general although there are some exceptions to them.

1. Most metals are lustrous. They are shiny, especially when polished.

2. Most metals are good conductors of heat and electricity at ordinary temperatures. Most nonmetals are insulators. There is also a variety of

materials that have the qualities of both conductors and insulators. Examples are carbon and silicon.

3. Most metals are malleable. They can be hammered or formed into sheets or otherwise shaped.

4. Most metals are flexible. They will bend under stress and return to their original shape when the tension is relieved.

5. Most metals are ductile. They can be drawn into wires or threads.

6. Most metals have tensile strength. They resist being pulled apart lengthwise.

7. Most metals are silver-white or grayish. Two familiar exceptions are copper and gold.

8. Most metal powders are dark gray or black. Notable exceptions are aluminum, beryllium, and magnesium, which are silvery.

9. Most metals are solid at ordinary temperatures. Of course, there is mercury, and a rare metal, gallium, which melts at 85°F. (29°C.).

10. Most metals will combine with oxygen to form oxides. The most costly oxide is formed by iron. Millions of dollars are spent each year for rust preventatives.

The names given to elements are supposed to reveal what they are. Recently discovered metals have names ending in *-um* or *-ium*, nonmetals in *-n* or *-ne*. Nonmetals conform to the rules pretty well, although sulfur, phosphorus, and arsenic go their own ways. Tellurium and selenium have metallic names although their properties do not always conform to what their names promise. A glaring exception is helium, an inert gas, which was discovered on the sun and was thought for a time to be the "sun metal." Such metallic nonconformists as manganese, cobalt, nickel, silver, lead, copper, zinc, iron, mercury, and antimony were named long before the rules were made up.

Hazards of Metals

Metals can be toxic or corrosive. No less than 20 metals are radioactive. And most metals have another hazard that is important to us: They will burn. The flammable potential of 81 different metallic elements varies about as widely as would the same number of hydrocarbons. For example, one of the reasons why precious metals are so prized is because they resist oxidation and thereby retain their luster. (If you doubt the value of this property, give the love of your life a ring of brass and note the comments when the brass

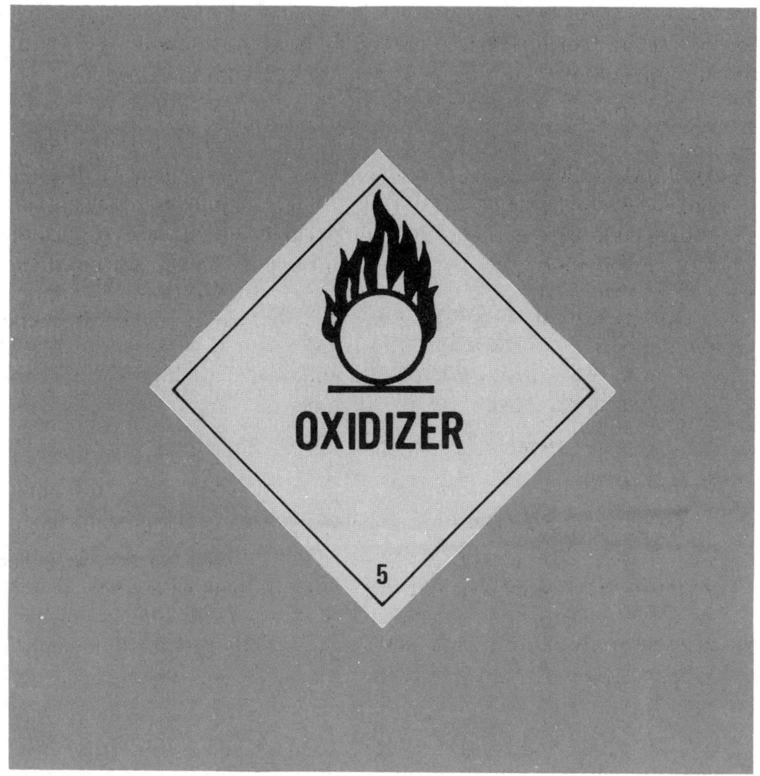

This yellow label with black lettering is required on containers in which oxidizing materials are shipped. A similar oxidizer placard is required when a shipment via motor vehicle or rail car carries a gross weight of 1000 pounds (454 kg) or more.

FIGURE 9–1. DOT Oxidizer Label

oxidizes and his/her finger turns green. A precious metal which steadfastly refuses to oxidize will also be stubborn about igniting readily.) Other metals are less resistant, and any metal which reacts with oxygen and releases heat is a candidate for ignition, if conditions are favorable.

Form and Shape

Perhaps the single most important condition which regulates the combustibility of a metal is its form and shape. In this respect, a metal is no different than any other combustible solid, be it charcoal, wood, or plastic. We will repeat this fact a dozen different ways during this chapter as we discuss the hazards of individual metals, but a few more general comments now may clarify the picture.

1. Some metals, difficult to ignite in a solid massive form, burn readily as thin sheets or shavings. Steel wool will burn, as will the aluminum foil ignited by an electric spark in a photographic flash bulb. As the division becomes finer and finer, the ignition temperature of the metal lowers. Surface areas grow and so does the hazard.

2. Dust clouds of such seemingly noncombustible metals as copper and tungsten are potentially explosive. What is true for these metals is even more likely for powdered titanium, a metal which burns readily. An industrial establishment which stores or creates metal powders or dusts by some process has the ingredients for an explosion on its premises.

3. Some metal powders can ignite spontaneously in air. When a material does this at ordinary temperatures, it is said to be PYROPHORIC. Pyrophoricity is a property shared by many flammable metal powders and dusts: magnesium, calcium, sodium, potassium, zirconium, hafnium, a special form of nickel, and several others. A layer of pyrophoric powder can ignite after a few minutes' exposure to atmospheric oxygen. It can ignite immediately after dispersion in air as a cloud. Metal powders have been known to burn in pure carbon dioxide, nitrogen, and argon, or even under water.

4 Some of these metal powders, when moist, are capable of producing an explosion more violent than one caused by nitroglycerin or TNT, and they can explode in the presence or the absence of air. What is most significant is that this heating, igniting, or exploding has been known to occur with no prior warning and no source of heat.

5. Many metals will react violently with water in one way or another no matter what their form, massive or finely divided. Some elements, the alkali metals (lithium, sodium, potassium, rubidium, and cesium), will explode when water comes in contact with them. They will be treated specially in the chapter on water-reactives in *Explosive and Toxic Hazardous Materials*. The metals we shall be concerned with here react most vigorously with water when they are heated or burning.

Combustion Environments

As we have mentioned, metals can burn in some surprising environments. But combination with oxygen is still the most common form of combustion.

Many metals are not too particular about where they obtain their oxygen. They will decompose water, burning with the aid of the oxygen in the H_2O molecule and releasing flammable hydrogen gas. Metals can also react with other chemicals to release hydrogen. Chromium, cobalt, iron, manganese, nickel, and zinc will react with acids; aluminum, titanium, and zirconium react with caustic alkalis.

The use of water streams on such metals merely intensifies the combustion process rather than cooling the situation. Metals have been known to burn more violently in contact with steam than with pure oxygen. Small pieces of many metals burn more intensely when wet than they do when dry.

The use of water on some metals has another hazardous possibility. Although some metals melt and burn at the same time (magnesium, for example), the dangers created by a molten metal are not necessarily confined to those metals that are flammable. When any metal contributes molten material to a fire situation, there is a good chance of a severe or even disastrous steam explosion if water is used unwisely.

We must consider fire-fighting methods carefully when metals, particularly metal powders, are involved. All of the common extinguishers—water, foam, dry chemical, carbon tetrachloride—and some of the inert gases may either stimulate the burning process or cause an explosion. A great deal of thought about alternatives and consequences must go into our sizeup and decision, especially if the fire also involves other types of combustibles, such as flammable liquids. The extinguishing agent recommended for a flammable liquid may cause a disaster if it meets a metal powder.

We will look more closely at specialized metal extinguishers later in this chapter, but even they are not entirely foolproof. Inconsistencies may pop up to bedevil us. Recommended methods of extinguishment may work beautifully on one metal fire only to fail in another, although the situation seems identical. The same metal from different sources can react differently to a particular extinguishing agent. A massive piece of pyrophoric metal can ignite at widely varying temperatures. One time it will ignite while resting on a piece of dry ice; another time it will fail to catch fire when heated to a melting point far above its normal ignition temperature. Much has still to be learned about the pyrophoricity and flammability of metals, even those as common as magnesium.

Magnesium

As far as industry is concerned, magnesium is one of the metals of the future. It is moderately hard and fairly strong. When alloyed with aluminum, zinc, and manganese, it is easy to machine and work. Most important, magnesium is very light, less than twice as heavy as water (1.74), only 65 percent as heavy as aluminum, 25 percent as heavy as iron. There are many uses for such a lightweight metal: in aircraft, automobiles,

TABLE 9–1. TEMPERATURES AT WHICH METALS BECOME MOLTEN

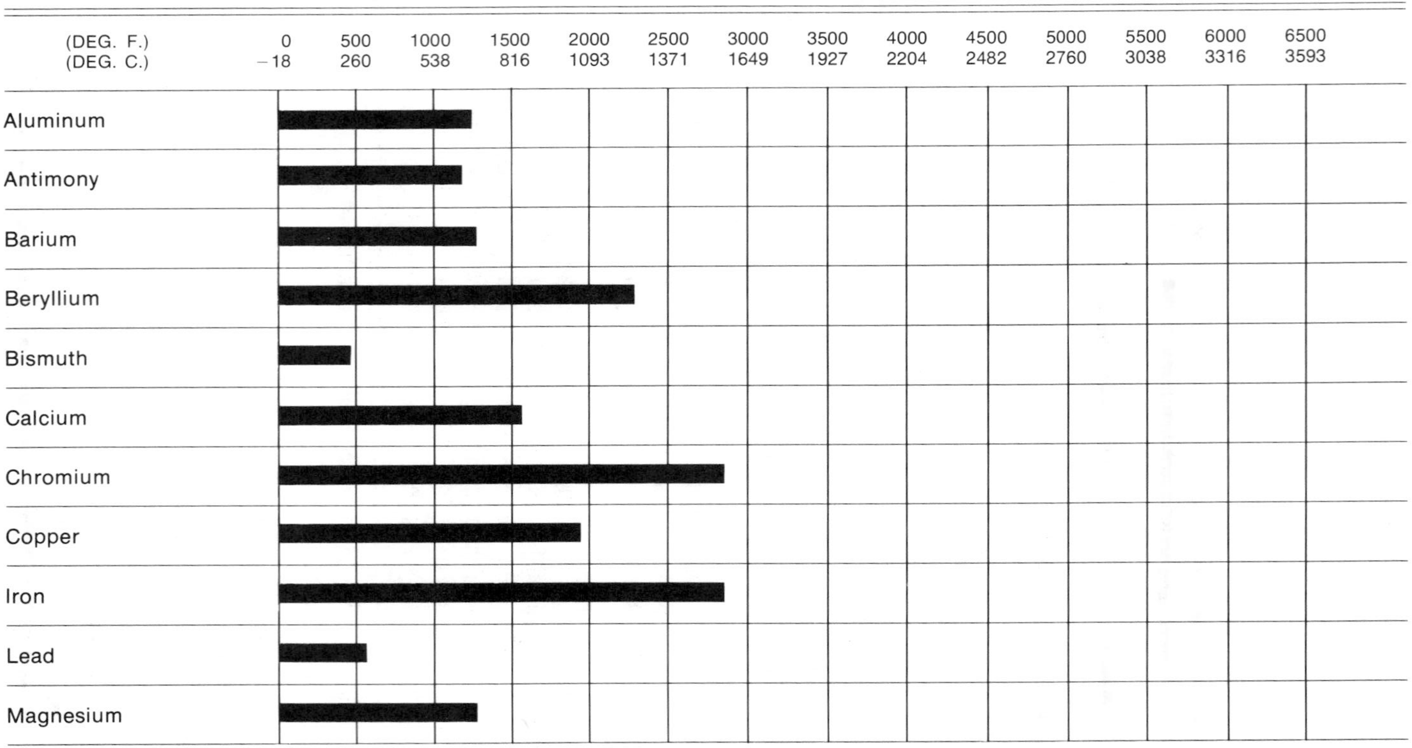

(DEG. F.)	0	500	1000	1500	2000	2500	3000	3500	4000	4500	5000	5500	6000	6500
(DEG. C.)	−18	260	538	816	1093	1371	1649	1927	2204	2482	2760	3038	3316	3593
Aluminum														
Antimony														
Barium														
Beryllium														
Bismuth														
Calcium														
Chromium														
Copper														
Iron														
Lead														
Magnesium														

TABLE 9–1. TEMPERATURES AT WHICH METALS BECOME MOLTEN (CONT.)

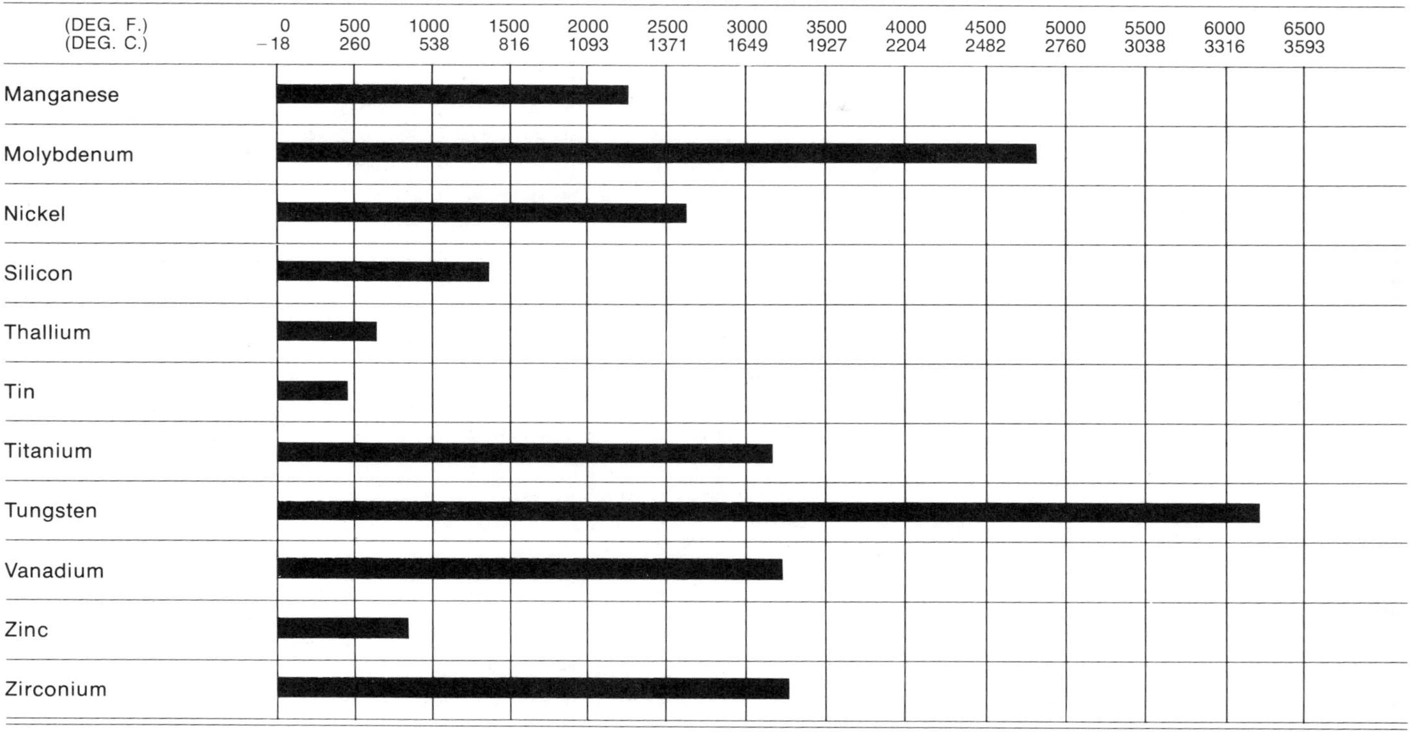

(DEG. F.) (DEG. C.)	0 −18	500 260	1000 538	1500 816	2000 1093	2500 1371	3000 1649	3500 1927	4000 2204	4500 2482	5000 2760	5500 3038	6000 3316	6500 3593
Manganese														
Molybdenum														
Nickel														
Silicon														
Thallium														
Tin														
Titanium														
Tungsten														
Vanadium														
Zinc														
Zirconium														

railway cars, buses, truck bodies, furniture, ladders, and luggage, to name a few. Magnesium alloys are even beginning to find some structural employment.

There is some reason for describing magnesium as the "silvery metal from the sea." Each cubic mile of ocean contains about 6.5 million tons of magnesium. While this is a lot of metal, the amount of water is far greater. Nevertheless, there is an economical extraction process. Seawater and slaked lime (often obtained from oyster shells) will yield magnesium hydroxide. This milk of magnesia reacts with hydrochloric acid to form magnesium chlorate, which, in turn, is split by electrolysis into pure magnesium and chlorine gas. The chlorine is recovered to make more hydrochloric acid. This method and more conventional mining produce an annual total of well over 200,000 tons.

Hazards of Magnesium

To the fire fighter, the future of the magnesium industry is less important than the problems the metal brings us now. Although inhalation of magnesium fumes or magnesium dust may cause some irritation of the throat and lungs, and the brilliant white light of burning magnesium has been known to hurt the eyes, it is not its toxicity but the flammable potential of magnesium that concerns us most. Uses of the metal and its powders reflect this property: pyrotechnics, photographic flash powders, military flares, and incendiary bombs. Magnesium, combined with white phosphorus in a bomb, makes a fearful device: the same water that controls the phosphorus causes flareups of the burning magnesium.

Magnesium melts at 1204°F. (651°C.) and boils at 2048°F. (1120°C.). The ignition temperature of a solid chunk is close to this melting point. Fine shavings and loose scrap will ignite at less than 1000°F. (538°C.); powder, another hundred degrees F. lower. Some magnesium alloys catch fire at less than 800°F. (427°C.). What is significant about these temperatures is that all of the common ignition sources reach these temperatures. This is especially true with respect to magnesium powder, whether deliberately produced for some purpose or unavoidably created by the grinding operations of a machine shop. As can be expected, heavier pieces of magnesium are far more difficult to ignite and are often stored in the open like other metals. As a wooden log will transmit heat away from a point source of ignition, so will a metal with a higher rate of heat conductivity. Although an entire piece of solid magnesium must be raised to its ignition temperature before it ignites, this is easily done in a fire. In addition, magnesium is one of those metals that melts as it burns. Larger pieces will contribute far more molten metal to the situation than an equivalent weight in smaller pieces because the smaller pieces are more likely to be completely consumed.

When heated, the magnesium molecule not only wants to combine with oxygen, it is positively fanatical about it. It will burn in pure carbon dioxide, combining with the O_2 in CO_2. While tests have shown that carbon dioxide does not accelerate burning a great deal, we expect something more from an effective extinguisher. Dry sand is also ineffective. Oxygen is jerked out of the silicon dioxide. If the sand is wet, the burning metal may produce steam and blow the sand pile apart. Burning magnesium will use asbestos as an oxidizing agent and continue burning even if the asbestos cuts off the air. The hunger for oxygen in burning magnesium is so great that it will burn more violently on a concrete floor than on a wooden one because the metal reacts with the oxygen in the concrete's aggregate. In spite of this, magnesium should not be stored on wooden floors.

Furthermore, if oxygen is not available, magnesium will ''make do'' with whatever is on hand. It will decompose carbon tetrachloride violently, getting at the chlorine atoms to support combustion. Even ''inert'' nitrogen gas supports combustion. Magnesium combines with it, forming magnesium nitride, and will burn in a pure nitrogen atmosphere. Argon, helium, and neon still remain aloof and will extinguish the burning metal.

Water on Magnesium Fires

In spite of what we have already mentioned, there is some disagreement over use of water on burning magnesium. The Magnesium Association in its bulletins to metal finishing shops states flatly: ''Never use water, fog spray, foam, common gas, or liquid fire extinguishers on magnesium.''

But other authorities, such as the AIA and the NFPA, partially disagree with this absolute ban on water.

> Magnesium, if hot enough, reacts violently with water with liberation of hydrogen, but the idea has been overdone that water should never be used on any kind of a magnesium fire.

When experts disagree, we must look for ourselves. We have two basic methods of applying water to a fire of any type: straight streams or fog patterns. Which method we will use on magnesium, if either, depends upon our sizeup. What amounts of metal are involved? What form is it in—powder, pieces, or large chunks? What is the extent of the fire? How long has it been burning? What is the type of building construction? How and where is the metal stored? What are the exposures?

If magnesium powder is present, it can be blown into the air by a straight stream. A dust explosion of tremendous intensity is possible. Straight streams must be used with extreme caution if magnesium dust is present. When there is a fire in a large quantity of small pieces, a straight stream can penetrate into the pile and cause an explosive scattering of molten and burning particles. The danger to personnel is evident. Although this scattering is often mistaken for a hydrogen explosion—the residue of the H_2O molecule after the oxygen is pirated away for combustion—most often it is simply a more intense version of an old-fashioned steam explosion, caused by the trapping of water.

If molten metal is involved, however, fog is much safer, for it will flash into steam immediately. But fog will not do the job properly. A small magnesium fire in chips and pieces simmers rather quietly if there is no moisture present. If water is applied, particularly water fog, the burning metal flares up amazingly into a hot, blinding white flame. Those who have seen experiments in which oxygen is deliberately applied to a burning combustible have seen a reasonable approximation of this intensive acceleration. Literally, the same thing is happening. Water is being decomposed by the ravenous desire of magnesium to combine with oxygen. Only large amounts of water can overcome this increase in burning rate and lower the temperature of the metal below its ignition temperature. This is particularly true when massive pieces are involved, the pieces most likely to produce molten metal.

To summarize, fog patterns are much more likely to accelerate the burning rate of magnesium, but there is less chance of a steam explosion. A straight stream can stir up an explosive dust cloud. There is more likelihood of a steam explosion if molten metal flows over pools of water we create or if water is trapped beneath the surface of a pile of chips. Yet, a stream produces more water in less time and we must have this volume of water to cool the metal below its ignition temperature and the temperature at which it combines so avidly with oxygen.

Modern methods call for a compromise between fog and straight streams, extinguishing and cooling the burning metal with a stream at lower pressures that will break up into drops over the fire. These coarse drops will not violently accelerate a magnesium fire the way fog will, but they will flow over the metal and cool it. After this, coordinated hose streams can be worked into the fire. Small but well-advanced magnesium fires have been extinguished by this method in a short time. If large quantities of metal are involved, it may be necessary to use large streams from a distance.

The problems of a big magnesium fire, in a finishing plant for magnesium castings, are described in AIA Bulletin 171, condensed below:

> The fire is of special interest as showing the value of hose streams on fires involving magnesium. A spark, from plugging in an electric drill, started the fire.
>
> Accumulated magnesium dust aided the fire's rapid spread to the wooden roof, filings and sawdust on the floor and shipping cartons.

Employees turned in the alarm promptly, tried to fight the fire with sand but were forced to flee. All 50 escaped unhurt.

The first companies of the Los Angeles Fire Department on the scene knew what was in the building from recent inspections; fire coming out the front windows and through roof skylights also had the brilliant white light of burning magnesium. To meet the need for many heavy streams, a second alarm was sent in at once.

Mounting a skillful attack, the firemen took hose lines into the adjoining sections of the building and onto the roof to stop the fire's spread; other streams, directly from the street, worked on the burning roof and on stocks of finished casting set afire by an early collapse of a portion of the roof. Twelve 2½-inch hose lines, with 1⅛- or 1¼-inch nozzles, brought the fire in the structure under control and then moved in from all sides to attack and extinguish the burning castings.

Water running off the roof and falling on burning magnesium produced minor puffs with some particle spattering but when heavy streams were applied directly to the burning magnesium, heavier explosions occurred. The early collapse of the roof, while offering a vent, allowed pieces of burning magnesium and metal containers to soar about 100 feet into the air. Windows were broken up to a block away, but no one was hurt. Pieces of burning magnesium, falling on the roofs of other buildings were quickly extinguished; many seemed to burn out before landing. Magnesium on the floor of the building, submerged in several inches of water bubbled but did not burn as long as it remained covered. The fire was under control in less than two hours and was confined to the section of the building where it started. About 8,500 pounds of castings, worth at least $2,300, were undamaged.

Failure to use heavy amounts of water would have resulted in a total loss of all the magnesium in the fire area and a probable spread of the fire to buildings across the street and the rest of the plant, with subsequent serious damage. Reports by the fire captains and chiefs who fought this fire are unanimous in saying only water applied in large quantity could have done the job. They also commented, however, that buildings for working magnesium should be of one story, with large window and skylight area, located outside of congested city districts. Magnesium explosions, in a more confined space, would have been much more severe.

Sprinklers

But why not prevent the formation of molten magnesium in the first place? Large pieces of the metal must burn for a time before molten pools will form. If extinguishment or control is immediate, these pools may not be created. An automatic sprinkler system above magnesium storage would be invaluable. The AIA and the Factory Insurance Association conducted a series of tests on magnesium dust, turnings, and small and large pieces with these thoughts in mind, and, as reported in AIA Bulletin 202, drew these conclusions:

> Automatic sprinkler protection is of definite value in the control of fires involving magnesium. Sprinklers, properly spaced and supplied with adequate water, should protect a structure from serious fire damage with burning of considerable quantities of magnesium therein. All tests and experience today show that water applied from sprinklers will not produce any reaction of sufficient violence to cause structural damage and therefore sprinklers should be allowed to operate until the fire reaches the quiescent stage. In certain other tests and in some fires, rapid evolution of steam and hot gases, capable of doing structural damage, have occurred when molten magnesium in quantity flowed into water which had accumulated to a depth of several inches from hose streams. Similar results might occur where conditions permit a quantity of water to flow into a mass of molten magnesium, but under automatic sprinkler protection the formation of molten magnesium would be lessened and with proper floor drainage the likelihood of such explosive effect would be greatly reduced.
>
> When the magnesium in a stack or pile becomes well ignited, it may be expected that a major part or all of the magnesium in that stack or pile will be consumed; automatic sprinkler protection can be expected to prevent the ignition of nearby stacks or piles of magnesium, as well as to reduce fire damage to other nearby equipment and materials.

Time can be of the essence in magnesium fires. Sprinkler systems can perform a vital function by holding down the extent of the fire until our arrival. If we can catch a magnesium fire while it is still small, several alternative courses of action may be available to us.

Dry Powders

Two types of dry powders have been approved for use on dry or oily magnesium chips, turnings, and castings. (The phrase "dry powder" refers to metal extinguishing agents. If a dry extinguisher is designed for such purposes as flammable liquids, it is officially called a "dry chemical.") One of them, PYRENE G-1, is a graphite powder which conducts heat away from the burning metal, with a possible additive which turns into a gas when heated and helps exclude oxygen. The other, ANSUL MET-L-X, is a salt (sodium chloride), with additives to prevent caking, combined with a plastic which melts and fuses the powder into a solid cake over the fire. Met-L-X will cling to vertical surfaces, a helpful property when larger pieces are involved. Both of these powders are noncombustible and nontoxic. They will not increase the combustion rate of magnesium if we run out prematurely. However, we must get close to fire before we can use them. The graphite must be applied by a hand scoop or shovel. The range of the sodium chloride extinguisher is less than 10 feet (3 meters), and care must be exercised not to blow burning metal around with the force of the ejected powder. If possible, good training should include practice with both of these powders.

Cooling water cannot be used on the magnesium fire in conjunction with these extinguishers. Not only would the flareup prevent close approach, but the water would destroy the cake we are trying to form over the burning metal. This cake should be an inch or so in thickness. Once it is in place, leave it alone. It will take some time for the metal to cool down.

Other possibilities exist for fighting a small magnesium fire. If the metal is on a combustible floor, we can put down a two-inch layer of powder, shovel the metal on top, and cover it with more powder. More sensibly, if a clean dry drum is available, we can make our magnesium sandwich inside it and carefully haul the drum to a safe place. A few burning chips can be quickly handled by dropping them in a bucket of water. (These dry powder extinguishers have not been approved for magnesium powder by the Underwriters' Laboratories.) If the metal powder is exposed to a fire, try to separate the two before the powder becomes heated. Be careful about the use of water. Remember that *damp* magnesium powder can heat, generate hydrogen, and possibly explode.

Storage and Use of Magnesium

The Magnesium Association makes several recommendations for safe machining and grinding. Keep their advice in mind when making inspections.

When machining magnesium, take heavy cuts with sharp tools, never permit a tool to rub on the work, and always back off the tool when the cut is finished. When taking fine cuts, use a mineral oil sealant, *not* a water-soluble oil. Sweep up chips frequently and store in covered metal containers.

When grinding magnesium, it is "magnesium only" for a grinder unless the wheels are thoroughly cleaned before changing metals. All dust must be carefully cleaned up and not allowed to accumulate on clothing, surrounding floor areas, beams or sills. Use a wet dust collector (the dust is caught by a water spray as soon as it is formed and settles as sludge) for extensive grinding. Use the collector for magnesium only. Remove sludge regularly, preferably at least once a day. Place in covered, vented containers and remove to a safe location for immediate disposal. Partially dry sludge is highly flammable. Dispose of the wet sludge by burning outdoors in an isolated area on a well-drained layer of fire brick, free from previous residue. Place dry combustible refuse, paper and wood, on top of a 3″ to 4″ layer (7 to 10 cm.) of sludge and ignite the paper in a safe manner to avoid burns from a possible hydrogen flash. Observe burning from a safe distance upwind.

Different forms of magnesium should be separated in storage. Finely divided pieces should be stored in noncombustible buildings with the finest in containers protected from sources of ignition, moisture, and contamination by halogens and acids. Large pieces may be stored out of doors.

Magnesium is shipped in all forms, from large ingots down to the powder in closed metal or cardboard containers. *The scrap and powder must carry a DOT red label: flammable solid, and the dangerous-when-wet label* (see Figure 9–2).

The label is blue with black lettering and symbols.

FIGURE 9–2. DOT Dangerous-When-Wet Label

Beryllium

Magnesium is the most common of a group of six elements called the alkaline earth metals. Other members of the family are beryllium, calcium, barium, strontium, and radium. Radium, of course, is radioactive.

Beryllium has many attractive properties that can be extremely useful. It resembles aluminum, not only as a silvery metal or powder, but also in many of its properties. Beryllium is the lightest hard metal known, much harder than magnesium, and its specific gravity is only slightly higher: 1.85. Copper and aluminum, when alloyed with beryllium, gain great elastic strength. Such an alloy can be bent back and forth thousands of times without breaking, the perfect metal for a lightweight spring.

Coupled with its strength and low weight, the high melting point of beryllium, 2400°F. (1316°C.), makes it useful in rocket nose cones. It takes a hot fire to melt a large solid piece. Although beryllium will oxidize in air, it is the least reactive of the alkaline earth metals. It is much less flammable than magnesium, and is even used to decrease the flammability of molten magnesium and aluminum. Active at high temperatures, beryllium resists oxidation at normal temperatures because it forms a skin of stable oxide. As with most metals, the powder can form explosive mixtures with air.

Beryllium Hazards

Beryllium is extremely toxic. If it enters the body through inhalation of dust or fumes, or through a break in the skin, fatal poisoning can result from a single short exposure to incredibly low concentrations of the element or its compounds. Symptoms of poisoning include: skin irritation, eye burns, ulcers which will not heal, coughing, shortness of breath, loss of appetite and weight. These symptoms can be delayed from three months up to six years. Death rate after an exposure is about 25 percent. For a time, fluorescent tubes contained a compound called beryllium zinc silicate. After World War II, this compound was replaced by another, far less toxic, but an ancient tube can still be around. Consider all obviously old tubes suspect and treat them like poison; they may well be.

Beryllium Powder

The powder is shipped under a DOT poison "B" label in steel and fiber drums. It should be kept away from air, acids, and moisture by storage in tight containers, preferably under an argon atmosphere. Fire-fighting procedures include the use of a dry powder. Water is not recommended. More important, personnel *must* wear complete protective equipment and self-contained masks. After a fire involving beryllium or one of its compounds, all personnel must bathe completely, with particular attention to fingernails and hair. Clothing should be washed separately from uncontaminated clothing. Equipment must be washed down thoroughly.

Calcium, Barium, and Strontium

Calcium is a silvery metal that rapidly tarnishes to bluish-gray in air. It is softer, lighter, and more reactive than magnesium. It will react rapidly with warm water to produce a stream of hydrogen bubbles, but this reaction generally does not produce enough heat to ignite the gas. (Calcium will detonate, however, on contact with alkaline hydroxides.) When finely divided, calcium will ignite in moist air and burn with a red flame. It burns rather quietly in air without creating molten metal. Its melting point is 1562°F. (850°C.). The burning metal produces highly irritating fumes of calcium oxide, better known as quicklime. The solid metal can react with the moisture of the skin and eyes to form calcium hydroxide, also caustic. Wear protective equipment and a self-contained mask; brush calcium dusts off equipment promptly. None of the common extinguishers are recommended. Water, foam, and halogenated hydrocarbons may cause an explosion or strong reaction. Dry chemical and carbon dioxide are inefficient. Met-L-X, Pyrene G-1, or soda ash are effective.

Calcium is shipped under a DOT red label as a flammable solid, a dangerous-when-wet label in airtight cans, well-stoppered bottles, and sealed drums. It is often stored under kerosene or naphtha. Containers should be protected from damage or high temperatures and kept away from areas where water may be used normally or in an emergency. An isolated, noncombustible building is recommended for quantity storage. Although many calcium compounds are common and useful, the pure metal has a limited employment in metallurgy, to harden lead, and in vacuum tubes.

Most of the fire properties, reactions, and storage requirements of barium and strontium are similar to calcium. Barium is silvery white. It melts at 1300°F. (704°C.). The pure element is not commercially important. All water-soluble barium compounds are poisonous.

Strontium is a soft, silver-white metal that turns pale yellow in air. It melts at 1386°F. (752°C.). There are no commercial uses for the pure metal. Both barium and strontium burn in air with a brilliant flame. Strontium chlorate, nitrate, and peroxide are all oxidizers and require DOT yellow labels with black lettering.

Titanium

The story of titanium is that of an ugly metal duckling turning into a valuable swan. For a long time after its discovery, titanium was just another laboratory curiosity, one of those silvery metals in the back pages of chemistry textbooks. Although it is one of the ten most abundant elements on earth, it is difficult to produce in quantity because it is spread about too evenly. It would have been more convenient to have it piled up someplace where it could be mined. Even when the metal was isolated, it had a disappointing property—extreme brittleness.

Then, a method using molten magnesium was developed which could prepare titanium economically in a really pure form. It turned out to be tough, malleable, shock-resistant, and so hard it scratched steel. The weight of titanium is halfway between iron and aluminum. But it is four times as strong as aluminum and stands up under heat much better. In fact, pound for pound, titanium is stronger than any metal, including steel. (Before you rush out and sell your steel stocks, remember that a rod of steel one inch in diameter is much stronger than the same size rod of titanium. Of course, it also weighs a lot more.) Titanium also proved to be resistant to many types of corrosion, including salt water.

Commercial production of titanium has increased enormously since World War II and the price has steadily dropped. There is every reason to expect that titanium will rank next to iron and aluminum as the most important structural metal of the future. It is used in jet engines, in marine equipment, including propellers and shafts, in machinery, orthopedic appliances, sporting equipment, and on and on. It is available in sheets, bars, tubes, rods, wires, sponge, and powder.

When we consider the other properties of this wonder metal, however, we find several drawbacks, particularly to the fire fighter. Compared to beryllium, titanium receives a clean bill of health as a poison. Its dust is considered only a nuisance. But, titanium is one of the metals that burns in massive form. Comparing it with magnesium, we see:

1. The ignition temperature of a titanium chunk in air is higher than that of magnesium, above 1300°F. (704°C.). Large solid pieces do not ignite and burn as readily. Short-time forging and fabricating of titanium alloys in temperatures up to 2200°F. (1204°C.) is possible. There have been some reports of spontaneous ignition of solid pieces of titanium subjected to jets of pure oxygen.

2. Titanium has a higher melting point than magnesium: 3100°F. (1704°C.) as compared to 1200°F. (649°C.). The heat of a surrounding fire is unlikely to reach temperatures this high, lowering the chances of creating quantities of molten metal.

3. While the flame temperature of magnesium is close to 2500°F. (1371°C.), titanium burns still hotter. Fine pieces burn slowly, producing great quantities of heat.

4. Strangely, although the solid metal is less combustible than magnesium, the reverse is true when titanium is reduced to a dark gray dust. The ignition temperature of magnesium dust is around 900°F. (482°C.). A layer of titanium dust can ignite or explode when exposed to temperatures between 720°F. and 950° F. (382°C. and 510°C.). Most common ignition sources are at least this hot. When this dust is thrown into the air as a cloud (perhaps by a careless hose stream), an

explosion is possible, given an ignition source between 630°F. and 1100°F. (332°C. and 510°C.). One of the reasons for this wide range is that titanium, produced by different processes, varies in its susceptibility to ignition. Titanium powder will decompose water and produce hydrogen. Powders immersed in or wet with water have been known to spontaneously ignite. Titanium powder ignites in pure carbon dioxide at 1260°F. (682°C.) and combines with nitrogen even more eagerly than magnesium at temperatures above 1475°F. (802°C.). The powder can also ignite upon impact, due to static electricity or friction, such as is produced by the grinding process. When titanium is being ground, it will produce flying sparks that leave a white trail, ending in several branches. This is unmistakable once seen. (See Table 9–2.)

Most of the recommendations for grinding and machining titanium, and disposal of its sludge are the same as for magnesium, although a more wary eye must be kept on accumulations of dust. There are no special storage requirements for massive pieces. For finely divided forms, which can also generate hydrogen when moist or oily, recommendations for segregated storage piles, good housekeeping, and covered metal containers are similar to those for magnesium.

Shipping of titanium powder is subject to several DOT regulations regarding containers. Whether wet or dry, the powder must carry a DOT red label: flammable solid. When shipped wet, with 20 percent water, it must be inside tightly closed metal cans not exceeding one gallon (4 liters) each, packed no more than twelve cans to a wooden box. When dry, titanium powder is shipped inside metal containers with soldered or screw caps. These cans must not weigh more than ten pounds (5 kg) each, and the wooden boxes containing them must not exceed a total of 75 pounds (34 kg). (The container must be cushioned by a layer of asbestos wool or rock wool. Dry powder, shipped in one-way metal drums, must also be protected by one of these noncombustible insulating wools.)

As with magnesium, all of the common extinguishers are either ineffective or downright dangerous. Although a coarse spray has been recommended for small quantities of fine pieces, water should be used cautiously. There have been violent reactions when water has been applied to hot or burning titanium. Many of the fire-fighting methods for magnesium are equally workable for titanium: isolation of burning materials from unburned, metal sandwiches; and the use of such specialized metal extinguishing powders as sodium chloride or graphite. Because the price of titanium is high, chips are salvaged by dropping them into volumes of water. Each pound accumulated is worth a few dollars.

Titanium has already made the "big time" in terms of fires. A 1958 fire involved 2 million pounds of scrap in a pile 300 feet long, 65 feet wide, and

TABLE 9–2. IGNITION TEMPERATURES OF METALLIC DUSTS

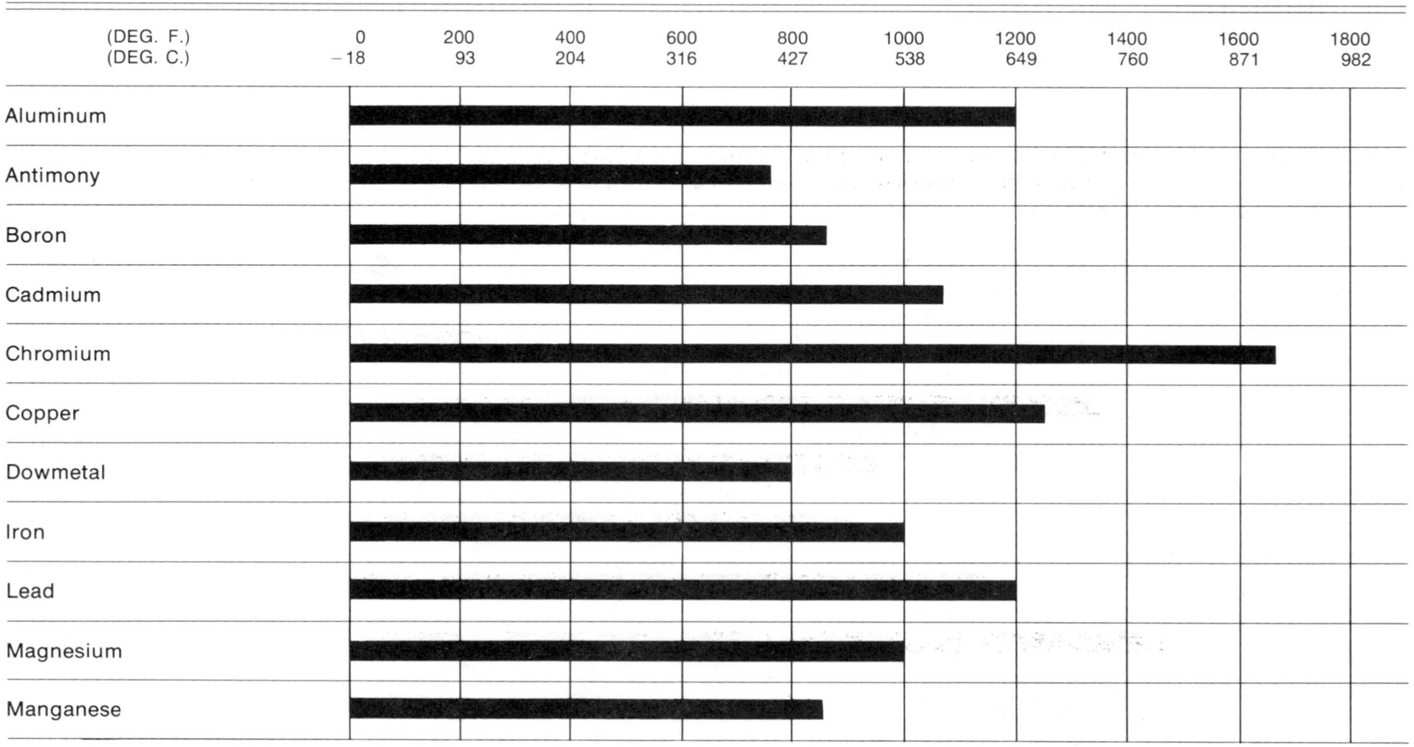

(DEG. F.) (DEG. C.)	0 −18	200 93	400 204	600 316	800 427	1000 538	1200 649	1400 760	1600 871	1800 982
Aluminum										
Antimony										
Boron										
Cadmium										
Chromium										
Copper										
Dowmetal										
Iron										
Lead										
Magnesium										
Manganese										

TABLE 9–2. IGNITION TEMPERATURES OF METALLIC DUSTS (CONT.)

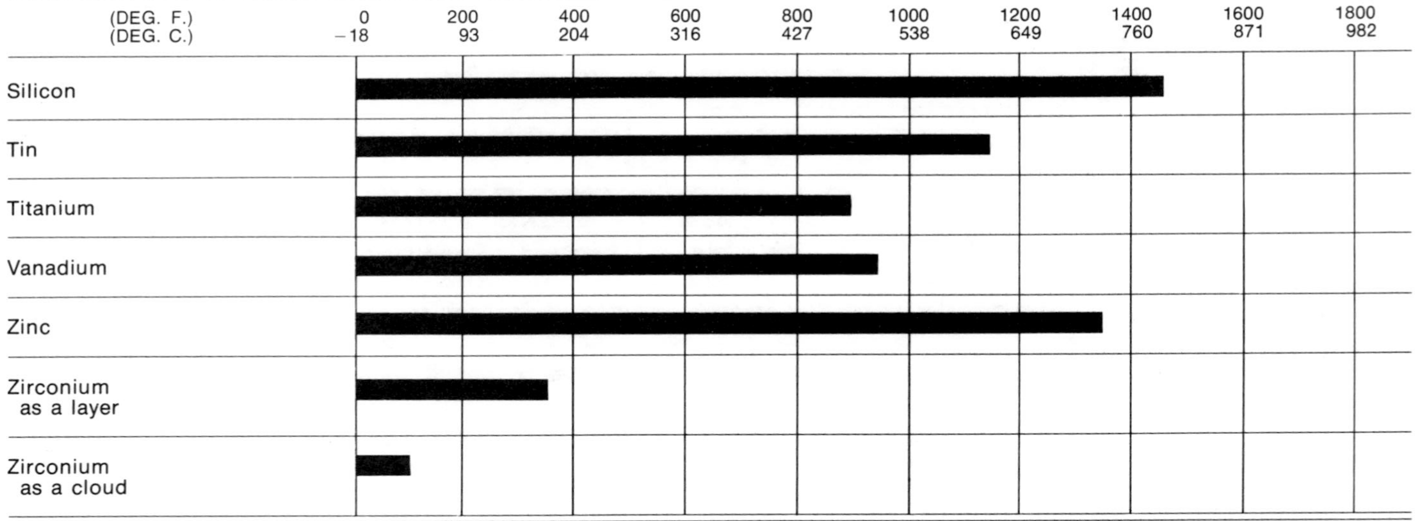

NOTE: A metallic dust can have varying ignition temperatures. Aluminum will ignite between 1195°F. (646°C.) and 1290°F. (699°C.); iron between 600°F. (316°C.) and 1435°F. (779°C.); lead between 1075°F. (579°C.) and 1310°F. (710°C.); magnesium between 970°F. (521°C.) and 1005°F. (541°C.).

45 feet high. The spread of fire was rapid and intense despite the quick use of hose streams.

Zirconium and Hafnium

The flammability of a massive piece of metal is not necessarily related to how it will perform as a powder. A chunk of magnesium burns more readily than a chunk of titanium, while the reverse is true when they are both reduced to dust. ZIRCONIUM goes a step beyond these two. There are no storage requirements for large pieces of the metal. Its melting point is 3326°F. (1830°C.), and there is little likelihood of the creation of molten metal. Unless somehow subjected to a jet of pure oxygen, massive pieces have not been known to ignite spontaneously. Indeed, a piece of zirconium at least ⅛ of an inch in all directions will not keep burning in air unless assisted by an outside heat source. But, there is an inverse relationship between the size of the individual particles of zirconium and its fire hazard; the danger grows greater as the size of the particles grows smaller. Chips are less dangerous than coarse dust; coarse dust less dangerous than fine dust. Very fine powder is *extremely dangerous*. A DOT red flammable solid label is required.

To evaluate the fire hazard, we must know the difference between the different kinds of finely-divided zirconium we have mentioned. We do this in the laboratory by measuring particle size, using the MICRON (about 1/25,000 of an inch) as our "yardstick." "Coarse" zirconium dust consists of particles measuring about 18 microns; "fine" zirconium powder consists of particles that measure just over three microns. In a fire situation, there is no practical way to tell these forms apart; we must rely on labels or correct information given to us by plant personnel.

Very fine zirconium dust, as a layer or a cloud, will ignite in carbon dioxide at temperatures between 1150°F. and 1200°F. (621°C. and 649°C.), or in nitrogen, at around 1450°F. (788°C.). In air, a layer of dust will ignite at 374°F. (190°C.). But we are most concerned when this fine dust disperses in air as a cloud. When this happens, the ignition temperature is only 68°F. (20°C.). This pyrophoricity can be due to frictional heat or generation of static electricity caused by grains of dust bumping against each other. For this reason, zirconium dust must *never* be poured through the air.

Factors which contribute to the flammability of fine zirconium powders include the depth of a pile and the moisture content. Damp powders, containing between 5 percent and 10 percent water, are considered those most likely to explode. They will burn much more vigorously than dry powder. Although hard to ignite, once started, zirconium dust burns more vigorously completely under water than it does in air. Disposal of fine zirconium powder has also proved very hazardous. We should attempt it only with the aid of competent technical advice. Scrap is generally ignited by a long fuse connected to a gasoline-soaked rag with the cans opened by a remote-controlled cleaver, and every precaution is taken against a premature ignition.

The DOT recognizes the importance of particle size in its shipping regulations. Special packaging requirements, including sealed metal or glass containers, cushioning materials, and weight limitations are required for zirconium powder or sponge, wet or dry. A yellow label, flammable solid, must be attached. However, zirconium powder exceeding 20-mesh* in particle size is not subject to these regulations. (See DOT Hazardous Materials Regulations, Section 173.214, for exact details of regulations.)

The working of zirconium metal is most often done under water or oil-coolant, with the aid of a wet dust collector. Scrap must be cleaned up frequently and no more than a quart of debris should be allowed to accumulate. The rate of combustion of zirconium chips and turnings varies with the size of the pile, how it is packed, and how wet it is. These fine pieces will not catch fire under water the way the dust will, and they will burn rather peacefully if they are dry and clean. But when wet, they burn with a brilliant white flame, eject burning material, and are likely to burn so fast no extinguishing method is possible. ''Dry powders'' such as graphite and sodium chloride will give control over a small fire in wet chips, and they are approved for extinguishment of oily chips. The application of water will intensify the combustion rate, although some tests have shown that, after a brief flare, a small quantity of chips can be extinguished by water. A few chips on fire can be drowned. For powder fires, prevent spread. Ring the fire with an approved metal extinguishing powder. Allow it to burn out. But do not create a cloud of fine dust; it will explode immediately.

Titanium, beryllium, and magnesium have been called the ''space age'' metals because they are widely used in air- and spacecraft. But this is also the age of the atom, and many metals have come into prominence because of the requirements of an atomic power plant. Zirconium and hafnium are two. Zirconium is unaffected by an atomic particle called a neutron and is used in the inner liners of nuclear reactors. Because the powder is highly flammable, it finds some use in flares, blasting caps, and photoflash bulbs. In addition, alloys of steel and zirconium are bullet-proof.

Although zirconium is more common in the earth's crust than nickel, copper, or tin (a cheap jewel called a zircon has a great deal of zirconium in it), U.S. production in 1945 amounted to only 20 pounds. Through use of the molten magnesium process, however, production of zirconium is now approaching 1 million pounds annually. It exists as a dark gray powder (loose or pressed into cakes); as a porous mass, called metal sponge; or as a

*This is a common measuring scale for large particles such as sand and gravel. ''Mesh'' means the size of the holes in the screen used; 20-mesh particles will not pass through a hole 1/20 of an inch in diameter.

silvery-gray, lustrous, and ductile metal, either in plates, strips, bars, scales, or flakes. Zirconium is resistant to corrosion by acids, alkalis, and sea-water. Nitric acid can convert the powder into a shock-sensitive explosive.

One of the reasons for the long delay in the use of pure zirconium metal is that it is always found in ores with hafnium. Of all the elements, these two are the most inseparable. The fire characteristics, storage, handling, shipping regulations, and fire fighting of hafnium are similar to those of zirconium.

Hafnium is 100 times scarcer than zirconium. Hafnium has a great appetite for neutrons. It will slow down the pace of reaction in an atomic pile. The reactor in the nuclear submarine, Nautilus, used hafnium control rods.

Aluminum

ALUMINUM is the most common metal in the earth's crust, and ranks second only to iron in world production and use, well over 9 million tons annually. (An American found a method of producing the pure metal by electrolysis a hundred years ago. It is still the only major structural metal not produced by the smelting of ore.) The metal has many useful properties.

1. Aluminum weighs one-third as much as iron, with a specific gravity of 2.7. Where weight-to-strength ratios are important, such as in utensils and appliances handled by women, or in the frames, engines, and propellers of aircraft, a high percentage of aluminum is still used in spite of the increasing challenge of titanium.

2. Aluminum foils illustrate the metal's malleability; its use in long-distance power lines illustrates its ductility. Pound for pound, aluminum is a better conductor of electricity than copper.

3. Aluminum powder is the same silvery color as the metal, one of the few metal powders not dark gray or black. Mixed with linseed oil, aluminum powder forms the metallic paint so familiar to fire fighters.

4. Aluminum reacts rapidly with atmospheric oxygen. Surprisingly, this property is valuable because the metal quickly forms a layer of aluminum oxide which protects it against further oxidation, one reason why aluminum siding is so popular.

5. Aluminum is essentially nontoxic, although the powder can irritate the eyes and respiratory tract.

Properties of aluminum that either limit its usefulness or are potentially hazardous include a melting point of 1220°F. (660°C.). The siding on house trailers, for example, can melt so rapidly in a fire it gives the impression of burning. Creation of molten aluminum is a definite possibility, especially when large pieces are involved. With a melting point this low, structural aluminum is quite likely to fail in a fire situation. When alloyed with other metals to increase its strength at ordinary temperatures, the melting point can be lowered still further. Heavy scrap, or bars, forgings, castings, and sheets are difficult to ignite. In bulk, finely divided aluminum does not ignite easily, but it will burn in air, CO_2, or nitrogen. Powdered aluminum, as a cloud in air, will ignite explosively at temperatures between 1193°F. and 1292°F. (645°C. and 700°C.). The explosion pressure can be as high as 100 psi (7 atmospheres). No ordinary building can withstand such an explosion without proper venting.

When bulk aluminum powder is involved, our major concern is to prevent the formation of a cloud. Not only will carbon dioxide and dry chemical be ineffective, but the force of their pressurized application will create the very cloud we are seeking to avoid. Carbon tetrachloride, applied to aluminum powder, has caused fatal explosions. Water will also stir up the dust and contribute moisture which could intensify the force of the explosion. Our only fire-fighting alternative is to isolate the fire by the careful application of fine dry sand or approved extinguishing powder. The burning aluminum powder will form a crust and eventually extinguish itself. In the meantime, the area containing the fire should be completely sealed to prevent cloud formation, and all ignition sources should be shut off. (This procedure is also suggested for control of other metal powder fires, such as magnesium.)

When aluminum powder is mixed with a flammable solvent, the use of carbon dioxide may extinguish the fire, although the powder can become exposed to air, form a cake, and re-ignite. Covering with fine dry sand or an approved extinguishing powder is again advisable.

Aluminum powders are used in explosives and in some bronze powders which then share the general hazard. When mixed with iron oxide, aluminum powder forms THERMITE, employed to weld heavy sections of metal. The temperature of burning thermite can reach 5000°F. (2760°C.).

Iron and Steel

In spite of increasing use of such metals as aluminum, magnesium, and titanium, iron and steel are still the skeleton of our civilization. Just as compounds are divided into organic and inorganic because of carbon content, metals and metallic alloys are either ferrous or nonferrous. (FERRUM is the Latin word for iron; the chemical symbol is Fe.) Most of it is used to make steel. In spite of the furor over steel price increases, it is still the cheapest of metals.

Pure iron is not often found. Iron is a silvery-white, very reactive metal. It will tarnish (oxidize; rust) in air or water. Iron and steel are easily ignited in the form of powders, dusts, or wools. Fourth of July sparklers are often made of iron powders on a metal stick, and they can be lit by a match flame. Unoxidized particles of iron and steel can spontaneously heat. A possible dust explosion hazard exists, since ignition temperatures of powders and dusts are between 600°F. and 1435°F. (316°C. and 779°C.). This wide temperature range reflects the varying compositions due to varying methods of manufacture. When hot, iron and steel dust can react with steam to release hydrogen. To fight a fire in these powders, use an approved metal extinguishing powder.

Some fires have been reported in large piles of steel turnings contaminated by oils. One such fire involved about 1300 tons of oily steel borings stored on an isolated concrete dock. The pile was taken apart by a "clam-shell" (a type of crane bucket), and the fire extinguished by continuous use of large amounts of carbon dioxide. Water was not used because of the confirmed explosive reaction with the heated metal.

When massive, iron or steel will not burn in ordinary fires; temperatures are simply not high enough. Since these metals melt somewhere around 2700°F. (1482°C.), they do not usually contribute any significant amounts of molten material.

The toxicity of iron powder is minimal, but eye disorders can result if particles are not removed promptly, and fumes from heated iron can cause some lung trouble. Nevertheless, iron has nothing like the toxic quality of beryllium and some other metals.

Steel-making

Without wandering too far afield, we should have some idea of how a basic industry like this operates.

Iron ore is heated in blast furnaces. The removed impurities are called SLAG. The remaining metal, 90 percent pure, is known as PIG IRON. Depending upon the way pig iron is cooled, suddenly or slowly, it becomes one or another of the cast irons. Some cast irons can withstand high temperatures without noticeable distortion. Others will break if suddenly heated and cooled. (This can happen when hose streams strike a cast iron column during a fire. Structural steels must be used for construction.) If the pig iron is remelted and further purified in a puddling furnace, it becomes WROUGHT IRON, softer but tougher than cast iron. The major difference between these two irons is in their carbon content. Wrought iron has less than 1.0 percent carbon, pig or cast iron from 2.0 to 5.0 percent. In between these percentages are the carbon steels, deliberately created according to specifications in open hearth, Bessemer, or electric furnaces.

Iron rusts; so does steel. Unlike aluminum oxide, which hangs on tight and prevents further atmospheric assaults, iron oxide continually flakes off and exposes a new surface to fresh oxidation. In time, a large piece of iron or steel will rust away.

Steel Alloying Metals

Special steels are made by alloying iron with other metals. Many of these metals find their principal industrial use (over 100 million tons each year) in the formation of these alloy steels. Some of these metals and their properties appear in alphabetical order below.

Chromium

CHROMIUM, alloyed with iron and nickel, forms stainless steel, which is not only rust-resistant, but has great strength. Chromium is a silvery-white, hard, and brittle metal which will take a brilliant polish. Automobile chrome is steel with a thin layer of nickel covered by chromium. Although chromium does not tarnish in air and resists oxidation, it burns in oxygen at high temperatures and decomposes water when red-hot. It melts at 2929°F. (1609°C.). Only a moderate dust explosion hazard, the ignition temperature of a cloud under favoring conditions is about 1650°F. (899°C.). Chromium will produce hydrogen gas from strong acids. It is essentially nontoxic. A strategic metal that must be imported into the United States, it is available in several forms.

Cobalt

COBALT is alloyed with iron to make permanent magnets. It is a steel-gray shining metal that is stable in air or water at ordinary temperatures, and has a low order of toxicity. Annual world production is about 20,000 tons.

Manganese

Ninety percent of the MANGANESE used by industry goes into the alloying of other metals. Manganese steels are tough and resist wear. They are used in safes, steam shovel teeth, and rock crushers. The pure metal is a silvery to reddish-gray color. It melts at 2246°F. (1230°C.). Manganese dust will ignite in air at 840°F. (449°C.), in pure nitrogen above 2000°F. (1093°C.), and burns with an intense white light. It will produce hydrogen in contact with strong acids, or when it reacts with water or steam at temperatures of 200°F. (93°C.). Chronic manganese poisoning and paralysis can result from continued inhalation of fumes or dusts. No DOT label is required.

Molybdenum

MOLYBDENUM steels have high tensile strength and heat resistance, ideal for high speed tools. Molybdenum is a silvery-white metal. The powder will react with steam at 1300°F. (704°C.), and is a slight dust-explosion hazard.

Nickel

NICKEL steels have great hardness and high tensile strength. They can be used in automotive parts or even armor plate. Nickel itself is a hard, silvery metal. It is not considered toxic or a serious fire hazard, although a dust explosion is possible. The annual world production of this widely used metal is well over 475,000 tons.

Silicon

SILICON steels are used in springs because of their flexibility. Silicon is a nonmetallic element, commonly found in two forms: a dark-brown amorphous powder that will ignite in air at 1425°F. (774°C.) and silvery leaflets that will not burn.

Tantalum

TANTALUM steels, used in tools, are very hard. At higher temperatures, the bluish metal or black powder becomes reactive. Although it has a low order of toxicity, it has caused some skin injuries.

Tellurium

A nonmetallic element, TELLURIUM is used in iron and stainless steel castings. The dark gray crystals or amorphous powder will melt at 846°F. (452°C.) and burn in air with a greenish-blue flame. There is a moderate dust explosion hazard. Probably toxic, tellurium can be absorbed through the skin. People who work around this element often have a breath which smells like garlic.

Tin

Tin cans are sheet steel covered by a thin layer of this silvery-white and ductile metal with a melting point of 450°F. (232°C.). In powder form, tin will oxidize in wet air. The dust cloud ignites at 1165°F. (629°C.). A solid piece of tin will decompose carbon dioxide at 1020°F. (549°C.), react with water vapor at 1200°F. (649°C.), and ignite and burn with a luminous white flame at 2370°F. (1299°C.).

Tungsten

TUNGSTEN steels are used in high speed tools because they retain their hardness when heated. Tungsten (wolfram) is a steel-gray metal with a high luster. The finely divided powder can be pyrophoric, and the metal reacts violently with various oxidizing agents.

Vanadium

VANADIUM steels, used in axles and gears, are shock-resistant. Vanadium is a steel-gray, extremely hard metal. The dust will ignite in air at 932°F. (500°C.). Prolonged exposure to concentrations of dust can lead to lung disorders.

Zinc

Besides alloying and painting, ''galvanizing,'' which gives metal a protective coating of ZINC, is another way to protect iron from oxidizing. Like aluminum, zinc protects itself with an oxide layer against atmospheric oxygen. A little less than half of the annual world production of 4.5 million tons of zinc is used for this purpose. Zinc is alloyed with copper to form brass and with copper and nickel to form German silver. It is also used in metal spraying.

Zinc is often called SPELTER in the metal trades. It is a bluish-white, malleable, and ductile metal and a bluish-gray powder. It is available in forms ranging from slabs to dust. In dust form, it is shipped in cartons, boxes, barrels, and drums.

Hazards of Zinc

The flammability hazard of zinc is a variation of a familiar story. Massive forms will not ignite easily, but they burn vigorously in air with a bluish-green flame, giving off quantities of white or bluish smoke once they catch fire. The metal melts at 786°F. (419°C.), a temperature low enough to create molten zinc readily in a fire. The dust forms explosive mixtures in air, with clouds igniting at 1110°F. (599°C.). Bulk dust, *when wet,* or when contacted by acids or lyes, will liberate hydrogen and may heat spontaneously to ignition. Powders must be stored in cool, dry areas, well-ventilated, and separated from acids, lyes, and moisture.

A fire in zinc powder should be smothered with an approved metal extinguishing powder. The use of water on massive zinc pieces is subject to all the hazards noted for other metals with low melting points, plus one addition: water, carbon dioxide, and halogenated hydrocarbons can react vigorously with burning zinc to form toxic gases. While the metal itself is not toxic, when heated or burning, zinc forms oxide fumes which are poisonous and cause ''fume fever'' or ''brass chills,'' characterized by a sweet taste, dryness in the throat, aches, chills, fever, and nausea. Fire fighters should wear self-contained masks to guard against these fumes.

Toxic Metals

Besides, beryllium and zinc, several industrially important metals add toxicity to their flammable hazard. The first of these is antimony.

Antimony

ANTIMONY (stibium) is silvery-white, hard, and brittle with a melting point of 1166°F. (630°C.). Although the pure element has no commercial use, antimony alloys and compounds are employed to harden lead in storage batteries and can be found in Babbitt metal (5 to 18 percent), in solder, printer's type, transistors, paints, ammunition, and glass. It is shipped as 55-pound (25-kg) bars or pigs, as lumps, or in finely-divided form. Antimony does not react with air at room temperatures but, when heated above 780°F. (416°C.), it ignites and burns with a bright bluish flame. When molten, it reacts with water to produce hydrogen gas and then reacts with the hydrogen to form an extremely toxic gas called STIBINE. This gas is sometimes given off by lead storage batteries. When heated, antimony can emit toxic fumes.

Bismuth

BISMUTH is hard and brittle, grayish-white with a reddish tinge, and has a bright luster. It is available in various forms. It burns in air with a blue flame. Although considered one of the least toxic of the heavy metals, it can cause kidney damage if taken into the body and should be considered a poison. Its melting point is 520°F. (271°C.). When 50 percent bismuth is combined with various percentages of lead, tin, and cadmium, it forms a fusible alloy that will melt at a specific temperature. These alloys form fusible links which open sprinkler heads and close fire doors.

Cadmium

CADMIUM is a grayish-white powder or a bluish-white, ductile, malleable metal soft enough to be cut with a knife. The metal tarnishes in moist air and burns when heated. Its melting point is 610°F. (321°C.). Cadmium has a wide range of uses. Cadmium plating of containers has caused food poisoning. *When heated, it emits highly toxic fumes.* A brief exposure to high concentrations can cause death from pulmonary swelling.

Lead

LEAD is a familiar, dull gray, very soft, heavy, highly malleable, and ductile metal. Over 3 million tons are produced annually. It is used for such purposes as cable coverings, plumbing, ammunition, gasoline additives, radiation shields, glass, printer's type, and storage batteries. Its melting point is 621°F. (327°C.). At high temperatures, it burns with a white flame. When at red-heat, rapid oxidation can be caused by air. The dust explosion hazard is slight; a cloud can explode between 1075°F. and 1310°F. (579°C. and 710°C.). Lead poisoning is one of the most common of occupational diseases. It is a cumulative poison, which means a little bit at a time can add up to a toxic dose because the body does not get rid of it easily. Poisoning can come from inhalation of dusts, fumes, or vapors, or sometimes through the skin. When inhaled, the symptoms develop quickly: anemia, pains in the joints and muscles, headaches, dizziness, and abdominal tenderness. If lead dust or heated lead is involved in a fire, fire fighters should wear protective equipment and a self-contained mask.

Mercury

MERCURY is the only metal that is a liquid at ordinary temperatures. It is a shiny, silvery, surprisingly heavy liquid. A 16-pound iron shotput or a lead ball will float on a pool of quicksilver. Shipping bottles of one, five, or ten pounds are quite small. Production is measured in 76-pound flasks, about 270,000 of which are produced annually. Mercury is used to extract gold and silver, in arc lamps, in explosives (fulminate of mercury), in thermometers, and, most commonly, in silent mercury switches.

Mercury melts at −38°F. (−39°C.) and boils at 676°F. (358°C.). These temperatures are important because mercury is a cumulative poison which can be taken into the body by inhalation, ingestion, or even through unbroken skin. Air, saturated by mercury vapor at 68°F. (20°C.), contains more than 100 times the toxic dose. Danger becomes even greater at higher temperatures as the boiling point is approached. Fire fighters must be careful to protect themselves against these vapors. Symptoms of mercury poisoning include excess salivation, pain upon chewing, nervous and mental disorders.

Thallium

THALLIUM is a bluish-gray metal which resembles lead. It is also toxic and should be carefully handled. Skin contact can be dangerous. One thallium compound is widely used as a rat poison. It decomposes water when red-hot, releasing hydrogen.

Glamour Metals

The remaining metals are the glamorous members of the tribe, the precious metals used in jewelry or those so rare there is only the slightest chance of meeting one in a fire situation. Still, strange things happen, and if a fire involves a precious metal, some of us are likely to be there.

Copper

The only inexpensive member of this group is COPPER, available in many forms and shapes. Copper is malleable and is one of the two metals which have a definite color. Hammered copper has been an ornamental metal throughout history. Since it is also ductile and an excellent conductor of electricity, much of the copper produced today goes into wire. For a long time, copper was second only to iron in industrial importance. Since the war, aluminum has taken over the second spot although 6 million tons of copper are used annually for ornaments, wires, machinery, piping, cooking utensils, and in alloys. Brass is copper (65 to 80 percent) and zinc. Bronze is copper (80 to 95 percent) and tin. Monel metal is two-thirds nickel and one-third copper. Where nonsparking materials are necessary, such as around flammable liquids, an alloy of copper and beryllium is often used.

Copper melts at 981°F. (527°C.). A dust cloud can ignite under favorable conditions at 1290°F. (699°C.). Although the pure element is low in toxicity, many copper compounds are poisonous.

Platinum Metals

OSMIUM is one of the platinum metals. It is white with a bluish cast, hard, and the heaviest of all elements. The dust will slowly oxidize at room temperatures. It is somewhat sparingly used in pen points or as a hardener for platinum. Although the metal is not highly toxic, upon heating in air, it gives off a pungent, nauseating, poisonous fume, osmium tetraoxide. Some of the other platinum metals have poisonous possibilities. All of them are expensive. Platinum itself has an irritating dust which may cause skin disorders.

RUTHENIUM dust ignites above 1300°F. (704°C.) and emits toxic fumes when heated. The other three platinum metals—RHODIUM, PALLADIUM, and IRIDIUM—as far as is known, have a low order of toxicity and slight fire hazards.

Rare Earths

One group of 15 elements is called the ''rare earths.'' Actually, they are not earths, but metals, and in a few cases, not particularly rare. However, most of them are uncommon and only require a brief mention.

CERIUM is the most abundant of these elements, more common than tin, silver, or mercury. It is a steel-gray, ductile, malleable metal, soft enough to be cut by a knife. This test is not advisable, because the pure metal has been known to catch fire when merely scratched. It will ignite above 300°F. (149°C.) and burn with a bright white flame. As with all metals, fine division increases the fire hazard. It melts at 1459°F. (793°C.). When heated, cerium will decompose water and liberate hydrogen gas. It is not considered highly toxic, although, like the other rare earths, little is really known about its toxicity. Misch metal, used in alloys and cigarette lighter flints, is made from a combination of rare earth metals, with over half of the alloy being cerium.

The other metals in this group are:

Dysprosium	Gadolinium	Lutetium	Samarium	Ytterbium
Erbium	Helmium	Neodymium	Terbium	Promethium
Europium	Lanthanum	Praseodymium	Thulium	(radioactive)

Most of these metals have properties similar to cerium. They will ignite in air between 300°F. and 350°F. (149°C. and 177°C.) (with the exception of lanthanum, known to ignite at 824°F. or 440°C.). They have varying rates of reaction with air and water. Many of them are stored in light mineral oil or sealed plastic. Two other metals, YTTRIUM and SCANDIUM, have closely allied properties. They are very expensive and rare, hardly out of the lab yet. The first pound of scandium was produced in 1960.

Remaining Metals

Eight other elements, many of them unfamiliar, complete the metals roster.

BORON is a black crystalline powder or soft brown amorphous powder which will ignite in air. It may cause poisoning if swallowed or inhaled. It is used to make Pyrex glass.

GALLIUM is a silvery metal which melts at 85°F. (29°C.) and boils at 3601°F. (1983°C.). The metal expands when it solidifies and can break glass containers. It is believed to have a low amount of toxicity.

INDIUM, a silvery-white metal, softer than lead, can be toxic if taken into the body. It melts at 313°F. (156°C.) and burns in air with a blue flame.

SILVER is the best metallic conductor of heat and electricity. Avoid inhaling dust. It resists oxidation but reacts with sulfur.

GOLD is a shiny, yellow metal, not toxic or acted upon by air or water.

GERMANIUM, a grayish-white metalloid, will burn in chlorine and bromine. It has a low order of toxicity, although some poisonings have occurred.

RHENIUM, a silvery metal or gray-to-black powder, is stable in air and very scarce. World supply is estimated at only 100 tons.

NIOBIUM, a silver-gray metal, reacts to oxygen and halogens only when heated and starts to oxidize at 392°F. (200°C.). Toxicity is unknown.

QUESTIONS ON CHAPTER 9

1. Name five properties shared by most metals.
2. Discuss the importance of the size and shape of a metal in determining its degree of combustibility.
3. What is the meaning of the world *pyrophoric?*
4. Give two reasons why the use of water can be dangerous around metal fires.

5. Discuss some of the sources of oxygen for burning magnesium.
6. What are the comparative merits and drawbacks of fog patterns and straight streams on magnesium fires?
7. Why are automatic sprinklers recommended for magnesium storage?
8. Describe two extinguishing powders that are recommended for some metal fires.
9. Name five metals that have a toxic hazard.
10. How do massive titanium and finely divided titanium compare in hazard with magnesium?
11. Discuss the hazards of zirconium powder.
12. What is our major concern in a fire involving aluminum powder?

BIBLIOGRAPHY

National Fire Protection Association:

No. 65, "Code for the Processing and Finishing of Aluminum."

No. 651, "Code for the Prevention of Dust Explosions in the Manufacture of Aluminum Powder."

No. 48, "Standard for the Storage, Handling, and Processing of Magnesium."

No. 652, "Code for Explosion and Fire Protection in Plants Producing or Handling Magnesium Powder or Dust."

No. 481, "Standard for the Production, Processing, Handling, and Storage of Titanium."

No. 482, "Guide for Fire and Explosion Prevention in Plants Producing and Handling Zirconium."

NFPA Quarterly, October 1957, "Pyrophoric Metals, a Technical Mystery."

Fireman's Magazine, May 1958, "Combustible Metal Fires."

Factory Mutual Association:

Handbook of Industrial Loss Prevention, Chapter 56, "Metals."

National Safety News:

"Dry Powders for Combustible Metals" (January 1963).

American Insurance Association:

Report: "Metalworking Plants, Fire Hazards and Safeguards."

Special Interest Bulletins:

No. 166, revised June 25, 1952, "Fire Department Operations, Hydrogen Explosions from the Decomposition of Water Under Fire Conditions."

No. 171, July 1, 1943, "Fire in Plant Engaged in Finishing Magnesium Castings."

No. 178, September 1956, "Magnesium."

No. 202, October 15, 1944, "Control of Magnesium Fires with Sprinklers."

No. 228, December 30, 1954, "Metalworking Plants, Fire Hazards and Safeguards."

VISUAL AIDS

"How Metals Behave," produced by the American Society of Metals, Pittsburgh, Pa., color, 30 minutes.

DEMONSTRATIONS

1. The combustibility of steel wool in oxygen is familiar, and easily demonstrated.
2. *Properties of Magnesium* (students and instructor should wear dark glasses): Ignite a magnesium strip to show the bright white light of the burning metal. If outside space is available, ignite a pile of magnesium "fines." When water is applied to the quietly burning pile, the flareup is something the student will not soon forget.

10 Plastics

Not satisfied with flammable liquids and gases or combustible solids, metallic and otherwise, American industry today is busily creating and using an enormous group of man-made materials that can also catch fire: the plastics. Defined in simplest terms, a plastic is a material that becomes moldable at least once because of chemical treatment, heat, or pressure, and subsequently hardens into a new shape.

It all began in 1833 when a French scientist discovered that if he treated cellulose, the basic component of plants, with nitric acid, he obtained a substance which burned cleanly and very rapidly. He called it XYLOIDIN. At that time, military gunpowders produced a puff of smoke that gave away the soldier's position. If xyloidin could be turned into a smokeless powder, it would be invaluable. It immediately became one of the first scientific babies to be drafted, but it proved difficult to train. Explosions occurred with distressing frequency as experimenters sought to apply xyloidin to military use. There was a noticeable lack of enthusiasm for continued research.

So matters stood for about 30 years, until concern began to be expressed about the fate of elephant herds being slaughtered for their ivory. A billiard ball manufacturer offered a $10,000 reward to anyone who could produce a substitute which would meet the exacting requirements of a cue ball. Tinkering with xyloidin, an American printer named Hyatt added camphor, which stabilized it enough to be worked. For awhile, from Jumbo's viewpoint, it seemed that xyloidin had left the battlefield for the pool hall in the nick of time. Alas, the new material proved too brittle for the fifteen numbered balls. But Hyatt had created the first usable plastic. In the years since its discovery, this mixture of xyloidin and a camphor PLASTICIZER* came to be known by its brand name, CELLULIUD.

Celluloid proved to have many excellent and economically useful properties. It was, it seemed, good for a lot of purposes outside the billiard parlor. It was resilient and tough, resistant to water and oils, could be produced in many colors and shapes, or buffed and polished to a high, attractive luster. Most important, it was cheap and easy to make. A whole generation of young men nearly choked to death in their celluloid collars. More and more uses were found for it: dental plates, piano keys, X-ray and photographic films, Ping-Pong balls, buttons and ornaments, shoe heels, toilet articles of various kinds, fountain pens, spectacle frames, toys, lacquers, paints, and patent leather. But as more was learned about its properties, Hyatt's invention came to be known by another name, PYROXYLIN, an appropriate word meaning "fire wood." For one disadvantage of this new material was a degree of fire hazard still unmatched by any other plastic. To find out why, let us back up a bit.

Cellulose Nitrate

The basic raw material, xyloidin, is known today as cellulose nitrate or nitrocellulose. It is produced by dipping some form of cellulose, generally wood pulp or cotton linters (small tufts of cotton stuck to the cottonseeds after ginning) into a solution of concentrated nitric and sulfuric acids. The process is called NITRATION. Any excess acid is removed by washing and boiling the fibers, which are then pulped before a final washing. The result is a white or near-white cottony fiber, chips, or powder. By varying the strength of the acids, the dip time, or the proportion of cellulose to acid, materials of differing properties are obtained. What nitration accomplishes is to attach nitrogen and free oxygen (ONO_2) to the cellulose molecule. The solubility, the uses, and even the names of the various forms of cellulose nitrate depend upon the percentage of nitrogen they contain. More significant to the fire service is that the higher this nitrogen percentage, the more potentially dangerous the resulting material.

Cellulose Nitrate Fire Hazards

The chemical formula for one form of cellulose nitrate is $C_{12}H_{17}(ONO_2)_3O_7$. Within the parentheses are three nitrogen atoms and *free, available oxygen*. Another form of nitrocellulose has four such groups attached. The more such oxygen in its makeup, the merrier any form of cellulose nitrate burns, even in the absence of air. Once ignited, the combustion process is very thorough.

*A plasticizer is any material added to improve the flexibility and other properties of a basic substance like xyloidin.

TABLE 10–1. TYPES OF NITROCELLULOSE

PERCENTAGE OF NITROGEN	NAMES	SOLUBILITY	USES	DOT LABELING
10.9–11.2	Pyroxylin Plastic, Soluble Nitrocellulose	Alcohols, esters, ketones	Paper coatings, plastics, low odor lacquers, printing inks	Flammable solid
11.3–11.7	Pyroxylin Plastic, Soluble Nitrocellulose	Alcohols, ether-alcohols, esters, ketones	Cellophane and paper coatings, alcohol-soluble lacquers, textile coatings	Flammable solid
11.8–12.2	Pyroxylin Plastic, Soluble Nitrocellulose, Collodion, Photocotton	Ether-alcohols, esters, ketones, glycols	Dopes, adhesives, coatings artificial leather, collodion, fast drying lacquers	Flammable Solid
12.6–12.8	Pyrocellulose, Pyrocollodion	Acetone	Propellants	Explosive A
13.0–13.8	Guncotton, Smokeless Powder	Acetone	Propellants, explosives, smokeless powder	Explosive A

It cannot smolder, for there are no places where oxygen cannot reach, since it is built in. Nitrocellulose is all surface. Methods of extinguishment based upon the exclusion of air are just a waste of time.

Pound for pound, cellulose nitrate does not produce as much heat as burning wood or paper. But the comparison is both misleading and meaningless, for it burns more than a dozen times faster. *A ton of nitrocellulose can be completely consumed in a little over a minute and a half.* The liberated heat is simply enormous, with flames extending like white blowtorches in all directions. As we shall see, spectators hundreds of feet away from a nitrocellulose fire have been injured. The quickness of the spread of such a fire is awesome and extremely dangerous. Because of burning rate alone, cellulose nitrate would qualify handily on most lists of extremely hazardous materials. Yet there is more trouble locked inside this molecule; nitration has made it unstable.

Like acetylene, cellulose nitrate will decompose, break apart, and release energy. But, while acetylene was primarily sensitive to *pressures* above 15 psig, nitrocellulose is sensitive to *heat*. Temperatures above 300°F. (149°C.) almost invariably cause decomposition; temperatures as low as 100°F. (38°C.) begin it. Such seemingly innocent sources of heat as steam pipes, friction, electric light bulbs, or even sunlight shining through

The labels for class A, B, and C explosives are orange with black lettering, and designated as Explosive A, Explosive B, or Explosive C.

FIGURE 10–1. DOT Explosive Label

unpainted windows are enough. Not being dependent upon the presence of air, decomposition is extraordinarily difficult to prevent. Once it starts, it is self-sustaining and constantly growing, heat-producing yet flameless. Soon the signs begin to appear—an acrid odor, a brownish discoloration, blistering, quantities of brownish-to-white gases. All the while the heat continues to build until, finally, the ignition temperature of nitrocellulose is reached. This is how most fires in cellulose nitrate storage are caused and one of the reasons why the addition of camphor is so important. It inhibits (discourages) the decomposition process. But aged stock can lose camphor through the normal evaporation of this volatile material. When it leaves, the remaining plastic becomes extremely vulnerable.

TABLE 10–2. GASES FORMED BY CELLULOSE NITRATE

GAS	FLAMMABLE?	TOXIC?	WATER SOLUBLE?
Carbon Monoxide	Yes	Highly	No
Hydrogen Cyanide	Yes	Highly	Yes
Nitrogen Oxides	No	Highly	No

Even the heat of decomposition does not complete the list of hazards. Decomposing cellulose nitrate produces gases in quantity. A single pound yields three cubic feet (.075 cubic meters) of gases, enough to make 400 cubic feet (10 cubic meters) of air explosive. Quantity storage can create enough gas to destroy walls, or even buildings, by the force of pressure alone. These gases are not only flammable and capable of exploding before or during a fire, but they are also highly toxic. Three of them which are of major concern to us are listed in Table 10–2. The flammability of these gases varies, as does their water solubility, which measures our ability—or a sprinkler system's—to wash them out of the air. However, all three are poisonous. The presence of the NITROGEN OXIDES is a final gift from the nitration process. These brownish-to-tan gases are insidious. They have little irritating effect on the respiratory passages, but once they hit the moisture of the lungs, they turn into nitric acid. Toxic, even fatal, symptoms can be delayed for several days. These gases are produced not only during decomposition, but also, in somewhat smaller amounts, when cellulose nitrate is burning, as AIA Bulletin 173 shows.

> It looked like only an ordinary fire, this blaze in a store handling supplies for students in one of the state colleges. Four of the firemen fighting the fire on the inside wore gas masks, but others were without them. Those without masks were dropping like flies. It was recognized then that the fumes were those from T-squares, triangles, protractors, and other engineering drawing materials made of pyroxylin plastic. All the men affected were immediately put to bed. One fireman completely collapsed; 45 minutes application of a resuscitator was necessary before normal breathing was obtained. Others were kept in bed, regardless of protests, until danger of collapse had passed.
>
> Fumes from pyroxylin plastic can sometimes be recognized by the smell of camphor; the smoke is dense, often of brownish color; the men who are gassed generally resist help with physical force. The fumes are very toxic and men without masks should not be allowed into the building. Water spray tends to wash out some of the toxic properties, therefore automatic sprinkler protection is very desirable.

The regulations regarding the transportation of cellulose nitrate depend upon the answers to many questions, among them:

1. How high is the percentage of nitrogen?

2. What is the physical form of the nitrate?

3. Has the nitrocellulose been dissolved, or is it moist or even dry? If moist, what kind of damping medium has been used?

A plasticizer, most often camphor, protects celluloid, although dry, from decomposing. Celluloid can be found in many sizes and shapes: sheets, rolls, tubes, rods, or as the finished article, a toy, an eyeglass frame, a toilet seat. Most manufactured articles are not under DOT shipping regulations Scrap celluloid, often susceptible to decomposition because of contamination, may also be shipped unless there are signs of decomposing. Then it must be under water. (The regulations for nitrate photographic and X-ray film will be considered later in this chapter when we compare it to Safety Base, cellulose acetate.)

Although the hazards of celluloid are more than enough to keep our undivided attention, cellulose nitrates with still higher percentages of nitrogen have also found widespread uses. Since the initial series of explosions which so discouraged the military, improved techniques have been found to tame xyloidin. Through the use of solvents, it can be softened into a dough which dries to a brittle solid suitable for making smokeless powder. Dry or even moist guncotton or smokeless powder can be exploded by a detonator. It was used as a high explosive for a time, but has been replaced by less treacherous materials such as dynamite and TNT. The high nitrogen forms of cellulose nitrates are considered to be explosives and subject to stringent shipping regulations. (We will explore this subject further in *Explosive and Toxic Hazardous Materials*.)

When nitrocellulose is totally dissolved in one of the suitable solvents mentioned in Table 10–1 (page 223), it is considered ''only'' as dangerous as the solvent itself. For example, when nitrocellulose is dissolved in a mixture of 70 percent ether and 24 percent alcohol, it forms COLLODION, a pale yellow syrupy liquid with an ether odor. Exposed to air in a thin layer, collodion gives a tough colorless film suitable for some medical and industrial uses. Collodion is shipped in glass bottles, in 1- to 10-pound cans (4.5 kg), and in 30- to 60-pound drums (14 to 28 kg) under a DOT red label: flammable liquid. All of these containers should be kept tightly closed when not in use. As we might guess, collodion has a flash point below zero.

Shipping and Storage

Whatever it may be called, industrial nitrocellulose with a lower percentage of nitrogen is generally shipped in steel 55-gallon (207-liter) drums, wet or dampened (not dissolved, mind you) with some sort of liquid, most often butyl alcohol, denatured ethyl alcohol, isopropyl alcohol, or water. When uniformly wet with 20 pounds (9 kg) of water to 80 pounds (36 kg) of dry material, cellulose nitrate carries a DOT red label: flammable solid. But often the final use of nitrocellulose does not permit the use of water as a damping medium. Shipped with one of the alcohols, the percentage of liquid required depends upon the physical form of the nitrate. If fibrous or cottony, at least 30 percent, by weight, of alcohol or solvent is necessary. When granular or flaked, it must contain 20 percent by weight. The flash point of the liquid used must not be lower than 30°F. (−1°C.). In all these cases, the drum carries a DOT red label: flammable liquid. There are also special limitations on the amounts which can be shipped by rail express.

Naturally, if the protection of the damping medium is lost, either through leakage due to physical damage to the drums or because the drums are not periodically inverted to assure an even distribution of moisture, the nitrocellulose will dry out. When this happens, the hazards are greatly increased. The presence of a flammable liquid also creates additional problems.

One of the largest manufacturers of nitrocellulose, E. I. DuPont de Nemours and Company, has several comments:

> Nitrocellulose is a flammable material which burns rapidly at a rate that varies with the kind and amount of damping medium. In addition to flammability, another principal hazard associated with the use of alcohol-wet nitrocellulose is the familiar one of volatile vapors. Explosive concentrations of air and alcohol vapor can accumulate in and around open nitrocellulose drums, especially if ventilation is poor, and such explosive "vapor pockets" are easily ignited by a spark, flame, or electric discharge. In the dry state, nitrocellulose is easily ignited by open flame, sparks, impact, or frictional heat resulting from pinching between hard objects. It burns extremely rapidly with a very hot flame, and does not require air for combustion. If it is ignited in a confined space, it will generate gas pressure with explosive violence. Nitrocellulose is sensitive to caustics and other alkaline materials. In direct contact with strong alkalis, nitrocellulose interacts with the evolution of heat, thus causing ignition of the material. Nitrocellulose is sensitive also to mineral acids.
>
> Unloading nitrocellulose from either a truck or a box car is a potentially hazardous operation unless it is well planned and executed with care and caution. It is difficult to specify a set of rules which will apply to all cases but rather each situation should be reviewed and planned fully to minimize the chance of a mishap occurring.
>
> When a rail or truck shipment is received, the door should be opened cautiously only part way, using a door opener on box cars, and a thorough inspection made to determine whether possible rough handling in transit may have resulted in spillage of nitrocellulose

> around the doorways. If nitrocellulose is observed on the floor, or around the edges of the doors, or particularly in the bottom track of box car doors, thoroughly wet the doors, edges, and floor with water before continuing to open them. After the doors are opened, inspect the interior of the car or truck for spilled drums. If loose nitrocellulose is observed, the entire interior of the vehicle should be wet down with water and, while still wet, the spilled nitrocellulose carefully cleaned up and immediately destroyed by burning. Thorough ventilation of the carrier is essential to purge from the car any alcohol vapors which may have accumulated; otherwise, there is the possibility of the formation of an explosive mixture of alcohol vapor and air.

Storage regulations for cellulose nitrate are based on two underlying principles. The first is fire prevention in grim earnest. All sources of heat, even those lukewarm, must be rigidly controlled. Water should be used as a coolant on machines, wire guards placed around common ignition sources, sparks kept to a minimum, stock protected against frictional heat, and an absolute prohibition enforced against smoking and matches. There should be periodic inspections for signs of decomposition, to make sure drums have not been physically damaged and that they are being inverted every 30 days. Of course, good housekeeping and all it entails is a necessity. Behind these fire prevention activities (and those mentioned in the standards listed at the conclusion of this chapter) must be knowledge of where this dangerous material is stored within our district, whether in film projection booths, X-ray vaults in hospitals, industrial establishments where paints, lacquers, or organic coatings are manufactured, or where articles of pyroxylin are stored or machined. A nitrocellulose fire should never come as a surprise to us. We should know where the potential exists.

The second principle: A large amount of material should not be allowed to decompose all at once. If a fire begins, it should involve only a relatively small amount. To this end, nitrocellulose should be segregated in storage. It should be stored in special isolated buildings, in vaults or cabinets, in work areas separated from one another by floor-to-ceiling partitions, or in special stock rooms of a fire-resistive construction. Even the use of tote boxes, wrappings, cartons, or paper envelopes to contain small amounts is helpful. It is curious, but logical once we consider it, that when cellulose nitrate has to burn *through* something, even a paper envelope, fire spread is greatly slowed.

This principle of segregation should be backed up by a building protected by decomposition and explosion venting and—very important—by a sprinkler system. Although sprinklers may not operate immediately because of the low initial temperatures of decomposition and although they may be overpowered by the resulting fire, nothing can replace a system with closely set heads, fed by a strong water supply.

Before we consider how to fight a fire involving some form of nitrocellulose, let us look at a couple of fire histories that show the intensity of the problem. The first is from AIA Special Interest Bulletin 1.

> Friday, June 9, 1933, was one of several very hot days. During the evening, many of the townsfolk of North Arlington, New Jersey, were finding relief by bathing in the Passaic River. Across the road from the bathers [was an] establishment devoted to the storage, sorting, and repacking of scrap pyroxylin. Suddenly, without warning, a burst of flame arose . . . to a great height and spread out in all directions; a lesser burst of flame followed shortly. The heat generated involved buildings and vegetation within a radius of 300 feet of the plant. Two adjacent buildings were involved, resulting in the death of six persons. Unable to escape the rain of burning pyroxylin, four of the bathers lost their lives, making a total of 10 fatalities. In addition, many were badly burned.
>
> The most tenable theory [of cause] is that of spontaneous heating to the point of decomposition . . . liberating large quantities of combustible and poisonous gases. That the fire involved such a considerable area is due to the fact that the gases were heated to their ignition point, ignition being retarded by the absence of sufficient oxygen within the building. When the pressure of the heated gases was sufficient to rupture a portion of the building and escape into the open air they ignited spontaneously. The pressure [also] served to propel them considerable distances, ignition occurring when the necessary mixture of gas and oxygen developed.

Another nitrocellulose fire, reported in the NFPA Quarterly (July 1961), occurred in a factory in Madrid, Spain, which made eyeglass frames. The saw, grinders, and other electrical equipment were located near the only exit. In the rear were the areas where people worked, the polishing room, and a stock room which contained a ton of nitrocellulose. The plant was designed for tragedy.

> Several times during the previous operation of this plant, sparks from the sawing operations had dropped into small piles of dust and chips. Previously, they had been confined and extinguished. At 6:30 p.m. on February 22, 1961, as the factory was about to close for the day, there was a large accumulation of dust and chips on the floor. A male employee was cutting a piece of the plastic on the saw when a burning chip jumped from the saw to the floor. The scrap ignited and flames quickly spread to the adjacent stock room. The blast of heat and smoke was practically instantaneous. The heat and dense smoke pushed against the concrete ceiling and wired-glass skylight.

Eventually, the glass melted, allowing the heat and smoke to start escaping, but by then the damage had been done.

Working in the factory at the time of the fire were 32 employees. These were mostly young girls between 16 and 22 years of age. When the sparks hit the floor, there was little time for anyone to do anything more than react instantaneously and instinctively. The 23 persons who found themselves cut off from the only exit pushed into the polishing room. Four crowded further into the adjacent storage room and several found a sheet of plywood which they pulled over themselves in a vain attempt to survive. They did not succeed, for the clouds of smoke contained, among other things, highly toxic oxides of nitrogen, one of the products of burning nitrocellulose. They died without a burn on them or their clothes—trapped for want of a way out. [Nine employees on the exit side of the fire managed to escape from the building. But five of them were burned, one so severely he later died.]

Fire Fighting of Cellulose Nitrate

We have seen the precautions necessary to prevent and lessen the consequences of a cellulose nitrate fire. Let us summarize now what such a fire can be like, what lessons can be learned from the two fire histories which have been included.

1. Burning or decomposing nitrocellulose produces toxic gases in great quantities. Even in a fire involving a small amount of pyroxylin, fire fighters have been hurt by these gases. Whenever cellulose nitrate is on fire or decomposing, fire fighters *must* wear self-contained masks. Every effort should be made to keep them upwind of a large fire. The area surrounding the fire should be evacuated.

2. Explosions can be caused in several ways:
   a. The amount of gas produced by burning or decomposition can cause enough pressure to force out walls.
   b. These gases are flammable. Produced in such quantity, they may pass quickly above the upper flammable limit when close to their point of origin. They will ignite or explode when they find an oxygen source (places where fire fighters are likely to be): at windows, in doorways, in adjacent rooms.
   c. Flammable liquids are often associated with nitrocellulose storage. Not only can their vapors ignite, but if a drum of nitrocellulose catches fire, the intense heat it will generate can cause a vapor pressure explosion in nearby drums, no matter what liquid is inside them, even water. Newly exposed, their

> contents burn in turn. A warehouse full of closely stored containers can become a gigantic, murderous, popcorn popper.

The two fire histories just given show the tremendous amount of heat produced by a nitrocellulose fire and how quickly it can spread. Too many people have been killed or injured attempting to extinguish a cellulose nitrate fire with insufficient water. Workers around this material should be warned to vacate the area if immediate fire-fighting measures are not successful. Tell them to leave it to the sprinkler system, which has no lungs, or to the professionals, who have learned to protect themselves. A cellulose nitrate fire is no place for amateurs.

And how do we professionals fight a cellulose nitrate fire? Every authority says the same thing: cautiously! There is only one extinguishing method open to us, for smothering is useless. Oceans of water, quickly applied to the entire surface, is the only method with a chance of succeeding: large amounts of water to combat the heat of combustion, quickly applied to combat the rapidity of the spread. The only fire-fighting appliance in a position to do this immediately is an automatic sprinkler system, if explosions have left it intact.

AIA Special Interest Bulletin 49 states the case as follows:

> To open doors or windows or enter a building where nitrocellulose is burning may be dangerous. In a recent nitrocellulose fire, two firemen were injured and one was killed by an explosion which occurred shortly after they had opened a door of a vault to direct a hose stream on the fire.
>
> The enormous amount of heat liberated by this material and the speed with which it burns makes it almost useless to attempt to extinguish it. This, coupled with the poisonous nature of the gases, produces conditions which materially alter the usual procedures of firefighting. When it is known that there is a large amount of film or of pyroxylin plastic articles and the odor of the smoke indicates that the fire has reached them, the men should be kept outside the building until the intensity of the fire has shown that the material is largely consumed. Protection of exposures is essential. Ventilation must be thorough before the men enter. Powerful streams from the outside are desirable, but the speed of burning often results in the material being consumed before these appliances can be put into operation. Never open the door of a vault, projection room, or storage room until fully assured all the material has been consumed, as an explosion is almost sure to occur when additional air is provided.
>
> Automatic sprinkler protection and adequate automatic venting are essential for storage of motion picture film and pyroxylin plastic.

> Without them there is a severe hazard to surrounding property and to persons nearby and to firemen.

The NFPA states flatly in Pamphlet No. 49:

> Use extreme caution in approaching fires involving this material as it may explode. No attempt should be made to fight advanced fires, except for remote activation of installed extinguishing equipment and/or with unmanned fixed turrets and hose nozzles.

And even when the fire is finally extinguished, our problems may not be over. There remains overhaul and possible disposal. Overhaul procedures must be carefully planned. Scrap pyroxylin after a fire may be extremely prone to decomposition after heating and possible contamination. If it is to be disposed of, it should be burned in the open, in small amounts, away from buildings, with due consideration of what lies downwind.

In spite of all its hazards, nitrocellulose is still popular in the United States if the number of trade names it bears is any indication. (See Table 10–3.) This is also true throughout the world. Many of the inexpensive sunglasses sold here are made in Europe from pyroxylin. Much publicity was given to the dolls from eastern Europe a few years ago which had celluloid heads. Nitrocellulose is widely used in the Orient because of its low cost. As toys, such finished products are often exported to this country.

But the unwelcome properties of cellulose nitrate, the fact that it discolors upon aging or in the sunlight, that it will dissolve in many common solvents, that it withstands heat poorly, and, above all, its hazards, led to a search for other plastic materials without these drawbacks. Nitrocellulose was the pioneer. It proved there was a ready market for man-made materials and showed the technology needed to create them. How well industry succeeded in finding substitutes for pyroxylin is the subject of the rest of this chapter.

How Plastics Are Made

The worst came first. Compared to cellulose nitrate, the hazards of other plastics seem minor. In most cases, plastics burn no faster than wood or paper. Their storage requirements, transportation regulations, and methods of extinguishment are similar to ordinary Class "A" materials. Nevertheless, they do have a few distinctive dangers of their own: the toxicity of their smoke, their methods of burning, and the hazardous company they keep.

Not only will fire fighters meet plastics in places where they are manufactured, fabricated or stored, but increasingly in every structure fire. Hundreds of pounds are in every modern home: in appliances, insulation, or even as part of the building itself.

Seemingly, there is a plastic for every use. What does your business need? A rigid or flexible container? A special piece of intricate shape? A

TABLE 10–3. SOME TRADE NAMES FOR CELLULOSE NITRATE

TRADE NAME	MANUFACTURER
Aceloid	American Cellulose Company; Indianapolis, Indiana
Amerith	Celanese Corporation of America; New York, N.Y. 10022
Celluloid	Celanese Corporation of America; New York, N.Y. 10022
Durakalf	Respro Incorporated; Cranston, Rhode Island 02910
Duralin	Respro Incorporated; Cranston, Rhode Island 02910
Fabrikoid	E. I. DuPont de Nemours & Co.; Wilmington, Delaware 19898
Febroid*	Textron Incorporated; Belleville, New Jersey
Gemlike	Gemloid Corporation; Elmhurst, New York
Gemloid CN	Gemloid Corporation; Elmhurst, New York
Herculoid	Hercules Powder Company; Wilmington, Delaware 19899
Hycoloid	Celluplastic Corporation; Newark, New Jersey
Inceloid N	American Products Manufacturing Co.; New Orleans, Louisiana.
Joda C/N	Joseph Davis Plastics Company; Arlington, New Jersey
Kodaloid	Eastman Kodak Company; Rochester, New York 14604
Macoid	Detroit Macoid Corporation; Detroit, Michigan 48204
Miracle	Miracle Adhesives Corporation; New York, N.Y. 10022
Multipruf	Elm Coated Fabrics Company; New York, N.Y. 10010
Nitron	Monsanto Chemical Company; Springfield, Mass. 01102
Nixon C/N	Nixon Nitration Works; Nixon, New Jersey
Plastite	Adhesives Plastics Mastic Company; Chicago, Illinois 60610
Polybond	Polymer Industries Incorporated; Sprindale, Connecticut
Pyraheel	E. I. DuPont de Nemours & Co.; Wilmington, Delaware 19898
Pyralin	E. I. DuPont de Nemours & Co.; Wilmington, Delaware 19898
Reskraf	Respro Incorporated; Cranston, Rhode Island 02910
Resyn	National Starch & Chemical Co.; New York, N.Y. 10017
Tan-O-Tex*	Columbia Mills Incorporated; Syracuse, New York
Terek	Athol Manufacturing Company; Athol, Massachusetts
Textileather	Textileather Corporation; Toledo, Ohio 43603
Texiloid	Textileather Corporation; Toledo, Ohio 43603
Tuflex	Respro Corporation; Cranston, Rhode Island 02910
Tufskin	Respro Corporation; Cranston, Rhode Island 02910
Wopaloid**	Worbla Limited; Papiermuhle-Bern, Switzerland

*Pyroxylin-coated fabric.
**Celluloid in sheets, tubes, and rods.

foam? A film? A fiber? A coating? A plastic is either ready to fill the bill immediately, or, if you wait a minute, one will be tailor-made to meet your specifications as to resistance to cold, heat, and chemicals, weight, strength, hardness, color, electrical conductivity, or whatever else you require.

Tailor-Made Plastics

Polyethylene

How to build a plastic to order? It all begins with a basic material like our old acquaintance, ethylene gas. As you remember, ethylene has a double bond between its carbon atoms (see Figure 10–2).

```
H   H
|   |
C = C
|   |
H   H
```

FIGURE 10–2. Ethylene (C_2H_4)

In a sense, a double bond is under some stress. If prodded, the carbon atoms will revert to single bondage. In the case of ethylene, this helping hand is provided by the application of heat and pressure (1,000 atmospheres) or through the action of catalysts. Either way, the double bond is broken (Figure 10–3). The unattached bonds immediately hook up with neighboring molecules. Another and another joins the lineup until chains 100,000 ethylenes long are formed.

```
   H   H
   |   |
 - C - C -
   |   |
   H   H
```

FIGURE 10–3. Rearranged Ethylene Molecule, the Ethyl Radical

Ethylene has become many (poly) ethylenes, polyethylene (Figure 10–4), the most widely used of all plastics. In doing so, it has changed from a gas to a waxy solid—not surprising, considering the similarity to a long-chain member of the paraffin series. The process by which these small molecules, or MONOMERS, join together is known as POLYMERIZATION. The gigantic molecules which monomers create are called POLYMERS.*

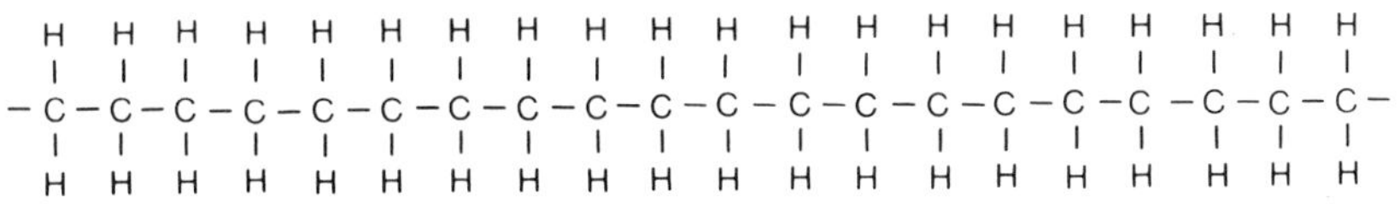

```
  H   H   H   H   H   H   H   H   H   H   H   H   H   H   H   H   H   H   H   H
  |   |   |   |   |   |   |   |   |   |   |   |   |   |   |   |   |   |   |   |
- C - C - C - C - C - C - C - C - C - C - C - C - C - C - C - C - C - C - C - C -
  |   |   |   |   |   |   |   |   |   |   |   |   |   |   |   |   |   |   |   |
  H   H   H   H   H   H   H   H   H   H   H   H   H   H   H   H   H   H   H   H
```

FIGURE 10–4. Polyethylene Chain

Polystyrene

Now let us try this process again with another monomer, for there is more to be told. When benzene and ethylene are chemically combined, they form a colorless flammable liquid with a strong aroma, styrene (Figure 10–5). The molecule resembles both its parents.

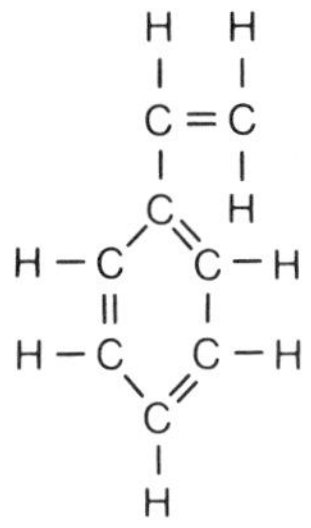

```
          H   H
          |   |
          C = C
          |   |
          C   H
        /  \\
    H - C    C - H
        ||   |
    H - C    C - H
         \  //
          C
          |
          H
```

FIGURE 10–5. Styrene ($C_6H_5CHCH_2$)

Styrene is much less stubborn about polymerizing. When uninhibited (unstabilized against polymerization) and allowed to sit at room temperature for several weeks, styrene will slowly turn into a thick liquid—a mixture of polymer and monomer—and finally to a glassy solid, polystyrene (Figure 10–6, page 236).

*Ethylene undergoes ADDITION polymerization, so-called for obvious reasons. Other plastics are made by monomers joining together after they squeeze out such smaller molecules as water. Nylon is made this way, by CONDENSATION polymerization. In addition, two or more monomers can be polymerized together to form a COPOLYMER.

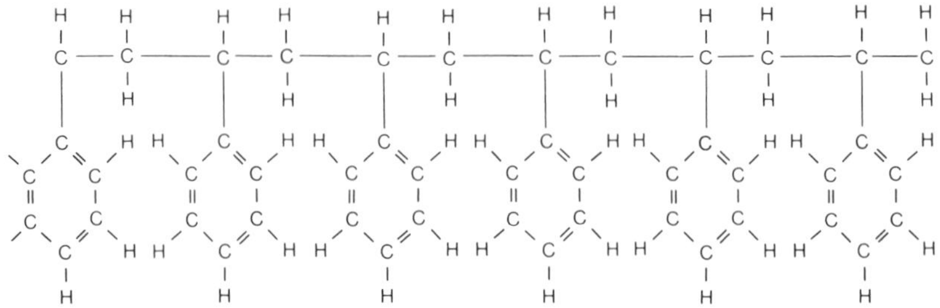

FIGURE 10-6. Polystyrene

If the surrounding temperature is raised to the boiling point of water, styrene will polymerize to polystyrene in a few days. Confined and raised to fire temperatures, the change can take place with violent rapidity. One of the major and continuing causes of explosions during plastics manufacture is uncontrolled polymerization, a reaction which runs away for one reason or another. As we shall see, some very dangerous chemical compounds are used to control polymerization, to promote and accelerate the process as and when desired.

Polyethylene and polystyrene polymers are basically thread-like. But other monomers can make linkages in all directions, forming a net-like structure. The difference between a net and a thread can cause many basic differences in the finished plastic. At some time in its life history, as its name implies, a plastic must be soft enough to be worked into a desired shape. A thread-like polymer can be softened repeatedly by heat, hardening into a new shape when cooled. In this, it is not unlike candle wax. It is called a THERMOPLASTIC. A net is more delicate and set in its ways. Some plastics can be given a permanent shape only once. Reheating will not soften them, for they will char and decompose if too much heat is applied. This type of plastic—for example, a polyester—is known as a THERMOSET. All plastics are one or the other, thermosets or thermoplastics.

So now we have our resin. What do we do with it? It all depends upon what the businessman wants.

Modern Plastics

A modern plastic may not be a single ingredient, but a blending of many substances into a compound.* The properties of a plastic can be altered greatly by the various additives which are thrown into the pot. BINDERS are another name for the resins we have been discussing. They can be synthetic in origin (polyethylene, polystyrene, polyester), natural (amber, shellac), or made from protein or cellulose. FILLERS can increase the bulk of the plastic at

*One of the phrases commonly used in the industry is "molding compound," for many plastics are molded into their desired shape.

low cost and also may improve the properties. They include such materials as wood flour, cotton fibers, metal wires, asbestos, clay, carbon blacks and so on. PLASTICIZERS are added to thermoplastics to develop such properties as flexibility. We have met one plasticizer already, camphor. Other additives can include solvents, lubricants, colorants, stabilizers, hardeners, catalysts, and flame retardants. From this witches' brew emerge plastics designed to meet many different and specific requirements. This is one of the reasons why plastics have become so popular. Requirements no longer have to be changed to meet the limitations of existing materials; the reverse can be true.

The Plastics Industry

Today, well over 6000 companies in the United States make plastics. A brief survey of the industry may ring a bell, for it is likely one or more of these companies are located within your city. They fall into three large categories that sometimes overlap: the plastic materials manufacturer who produces the basic resin or compound; the processor who converts the plastic into solid shape; the fabricator and finisher who further fashions and decorates the plastic.

The primary function of the materials company is the formulation of the plastic from basic chemicals. This plastic compound is sold in the form of granules, powder, pellets, flakes, and liquid resins or solutions for processing into finished products. Some plastics materials companies may go a step further and form the resin into sheets, rods, tubes, and film. Some purchase chemicals from which they formulate the plastic resins and compounds. Some only make the compound, purchasing the resin.

From the manufacturer, the resin or compound moves to the processer. Molders produce finished products by forming the MOLDING COMPOUND into desired shapes. Extruders may be divided into two groups. The first turns out sheets, film, special shapes, rods, tubing, pipes, wire coverings. The second includes producers of plastic filaments which can be woven into clothing, rugs, or screening. Film and sheeting processers work by calendering, casting, or extruding the plastic. High-pressure laminators form sheets, rods, and tubes from paper, wood, and cloth which are impregnated with resins. The reinforced plastic manufacturers take liquid resins and combine them with such reinforcements as glass fibers, asbestos, and synthetic fibers to form strong, rigid structural plastic. Finally, the coaters make use of various processes to coat fabric and paper with plastic.

Plastic sheets, rods, tubes, and special shapes are the principal forms with which fabricators and finishers work. Using all types of machine tools, they complete the conversion of these plastics into the finished product. For example, working with rigid sheet material, fabricators form such things as airplane canopies and television lenses. Plastic sheeting and film can be made into such products as shower curtains and rainwear. Many companies are engaged in printing, embossing, metal-plating or otherwise decorating the plastic.*

Hazards of Plastics Manufacture

A survey of the hazards of the plastics business divides itself into two parts rather neatly:

1. The multiple problems which surround the creation of a plastic.
2. The properties of the finished product.

First things first, for here lies the greatest likelihood of a fire or explosion. The basic materials—such monomers-to-be as butadiene, styrene, acrylonitrile, phenol, formaldehyde, propylene, ethylene, and the others—may have any combination of reactivity, toxicity, or flammability; some may have more than one or all of these properties. Many of these basic materials must be inhibited to prevent the polymerization which can produce heat and, possibly, explode. Acrylonitrile and phenol are only two of the many monomers which are poisonous. Some plastics, such as the cellulosics, are produced by the action of corrosives. Both flammable liquids and flammable gases are well represented, and with all this present, things can sometimes go wrong.

> In Louisiana, a $9 million fire occurred in a large petrochemical plant, consisting of a group layout of widely separated block-area units with open air construction, processing basic materials for the production of polyethylene and vinyl chloride resins. Failure of equipment caused a major spillage of ethylene, and the escaping vapor engulfed one of the area units. Within 30 seconds, the ethylene ignited with an explosive roar and quickly involved the process units—the ignition was believed to have originated at cracking furnaces 200 feet distant. The deluge (sprinklers) water system was rendered inoperable by the explosion. The fire was finally brought under control by use of hand hose and portable deluge sets, after 3 hours. Practically the entire structure of the process area was collapsed by the intense burning of 250,000 pounds of the product. This operation sustained a shutdown for several months for equipment replacement and structure rebuilding.**

*Information in the preceding four paragraphs is taken from "Plastics, the Story of an Industry," a booklet produced by the Society of Plastics Industry, Inc. Those interested in the actual processes of molding, calendering, casting, laminating, reinforcing, extrusion, and fabricating can find an authoritative summary there. Many of the facts on individual plastics are also taken from this source.

**Fire histories in this section are taken from AIA, "Fire Hazards of the Plastics Manufacturing and Fabricating Industries," 1963.

[In another fire] styrene polymer residue, coating the inside supports and roof structure of an empty styrene monomer storage tank of approximately a million gallons capacity, caught on fire as workmen were using welding and cutting torches during a cleaning and altering operation. In approximately 5 minutes (upon the arrival of the fire company) an explosion ripped a part of the cone-shaped roof from one side of the tank. Inaccessibility to the tank interior because of a 30-foot high steel dike limited firefighting to cooling the tank walls while the fire raging inside burned itself out in about 30 minutes. It was theorized that at welding temperatures, about 2500°F., the styrene polymer accumulations undergo pyrolytic (heat-loosening) decomposition, with some depolymerization and reversion to monomers, which, in turn undergo "cracking" to lighter flammable hydrocarbons. The formation of such hydrocarbons may have been responsible for the explosion.

Monomers are not alone in their combustibility. The entire flaming family is around when a plastic is born. Not only gases and liquids, but solids. Solidified resin is often ground or pulverized into finely divided form before use. The fillers previously mentioned include a number of combustible powders, and, as we shall see shortly, many finished plastics are explosive when reduced to dust during processing or fabricating. Flammable liquids are used as solvents in making plastic paints, adhesives, or coatings. Lubricants that reduce stickiness in molding processes are often flammable, especially if the mold is still hot. This list could be extended, but a fire department inspection of each plastics company in your area is a better way to complete it, for a number of fires have been caused by each of these possible sources.

A $4 million fire in New York was caused when hot resin was discharged into a mixing tank containing 900 gallons of naphtha. The naphtha was vaporized, and it escaped into the building and exploded. Two men were killed.

[In another large fire, this time in New Jersey and costing $8 million] the fire started in the resinous dusts accumulated on the wooden planking over a plastic laminating machine. Employees attempted to extinguish the fire with portable extinguishers. There was a delay in sending in the alarm and the fire spread rapidly between the walls and the flooring of the building. The sprinkler was only partially effective due to lack of water.

[In Massachusetts] a minor explosion in the vicinity of a polystyrene grinding unit was followed by a larger blast in the manner characteristic of a dust explosion. One man was killed and two others

> were badly burned. It was indicated that the dust housekeeping conditions in the plant were not satisfactory. The plant operated on a round-the-clock schedule.

The initiators or catalysts used to control the polymerization process include some dangerously reactive and highly flammable chemicals, the ORGANIC PEROXIDES. Among them is a powder which has a burning rate faster than cellulose nitrate. This is UNINHIBITED BENZOYL PEROXIDE. A more thorough examination of these peroxides, which have an annual production of over 15 million pounds, will be found in *Explosive and Toxic Hazardous Materials*. (Many of the monomers will also be found there.)

This fire history indicates their reactivity.

> Fire broke out in the mixing room of a reinforced plastic boat manufacturing plant. The fire originated in a refuse barrel and was probably caused by a reactive mix which contained organic peroxides. Employees were unsuccessful in attacking the blaze with their extinguishers before the volunteer fire department was summoned. The lack of water rendered the fire department operation ineffective. Loss: $296,000.

The industrial employment of the polymerization process generally takes place in kettles, or autoclaves, where temperatures and pressures may be as high as 1800°F. and 700 psi (982°C. and 48 atmospheres). (Molding pressures can run to 30,000 psi [over 2040 atmospheres].) Finagle's Law should certainly work here: Whatever can go wrong, will. Inadequate controls over any factor may cause the process to "run away": temperature, pressure, reaction times, too much catalyst added too quickly, the improper operation of relief valves, the premature removal of an inert atmosphere. When all this is piled on top of the ordinary fire prevention problems of every industry—improper housekeeping, the presence of unguarded sources of ignition, improper disposal of flammable materials (unless watched, some plants make a habit of using public sewers as a private waste disposal)—we can readily see why fires are not unknown in the plastics industry.

Two more fire histories:

> A violent reaction occurred in a 750-gallon reactor during the polymerization of vinyl acetate. Employees tried to control the runaway reaction by cooling but they were not successful. Observation ports of the reactor burst, releasing acetone vapors into the building. Ignition of the vapors caused a violent explosion. One employee was killed. It was reported that the usual pressure relief system did not operate because a 175 psi rupture disc was used instead of a 75 psi disc. Loss: $390,000.

> Contents of an electrically-heated resin kettle caught on fire upon heating the mixture, and subsequently boiled over, spreading the fire to an adjacent warehouse in which a large quantity of resin was stored. Some 250 firemen spent about 6 hours combating and limiting the fire to the process area, where the fire originated, and to the warehouse. Most of the effort was directed to protecting the exposed tanks and equipment, thus preventing the entire plant from being destroyed. Loss: $700,000.

And something new is being added, the increased use of RADIATION to form cross-linkages, to weave thread-like polymers into nets and improve such properties as the flow rate and tensile strength of a particular plastic. Radiation is also being employed to bind monomers permanently to construction materials.

Structural Plastics

With billions of pounds of plastics being produced annually in the United States, the hazards of the finished product are of more than passing concern to the fire service. After plastics leave the manufacturer, processor, and fabricator, they are bound to be piled up somewhere in a warehouse. Fortunately the polymerization process has removed much of the toxicity, instability, and flammability of the monomers. Yet, a fire in a warehouse storing several millions of dollars worth of plastic products has often resulted in a total loss. Let us find out why.

Plastics are rapidly taking their place alongside wood, glass, stone, and steel as major structural materials. The amount of plastic in a modern building can only be appreciated when we make a list: furnishings, cushions, padding, appliances, utensils, rugs, draperies and fabrics, veneer, siding, sealants, insulation, ceiling tiles, floor tiles, wall tiles, cabinetry, decorative panels, door and window trim, lighting diffusers, luminescent panels, window panes, paints and varnishes, structural panels, electrical fixtures, conduits, pipe, wire insulation. . . . In these applications, a dozen or more plastics can be employed. This is why we seek the answers to such questions as, How easily will these plastics ignite and sustain combustion? What type of gases and smoke do they develop? How much heat will they generate when burning? How rapidly will a flame spread across their surface? Even if not on fire, how will they react to fire temperatures?

Let us consider this last question, for it is more important than it appears at first glance. If there is one weakness shared by most plastics, it is inability to withstand continuous temperatures much above 300°F. (149°C.) and still stay in service. As structural materials, their thermal distortion is of direct concern. What happens when they soften or melt or deform at fire temperatures?

The plastics industry is well aware of the problem, recognizing their safety responsibility. Maximum service temperatures are invariably included in the specifications of a plastic (see Table 10–4). Only a very few are above 500°F. (260°C.)—some silicones and cold-molded materials, a fluorocarbon or two. But research is continuing. It has been discovered, for example, that the irradiation of polyethylene will greatly increase its thermal stability. Eventually a solution will be found. Until then, we must recognize that unprotected plastic will generally deform in a fire.

Representatives of the industry sometimes take a positive viewpoint. They point out that light panels, for example, will buckle and fall from their mountings before ignition, thus reducing the possibility of a flame continuing along the ceiling surface. They also argue that a plastic, used in windows, will distort when heated and provide a vent. Other authorities are less enthusiastic about falling sheets of plastic and sudden venting.

The AIA, in "Fire Hazards of the Plastics Manufacturing and Fabricating Industries," speaks glumly of the "costly replacement" of all those deformed panels and calls attention to another structural problem.

> Plastics have a tendency to "creep" when moderately stressed under an elevated temperature and some undergo irreversible deformation. For reliable use in design work, data on creep measurements should be obtained over an extended period of time. This characteristic is important for any plastic product under continuous loads such as pipes, pressure vessels, bearings, and structural elements. Several failures of reinforced polyester tanks in acid service at elevated temperatures have caused a more careful consideration of the entire subject of design and selection of materials.

Hazards of Finished Plastics

Most plastics will burn. Everyone agrees on that. But, always excepting cellulose nitrate, the *degree* of the flammability hazard of plastics is still under active discussion and investigation. Some authorities point to the fact that the self-ignition temperatures of most plastics are hundreds of degrees higher than those of ordinary combustible materials.

Since a solid material generally must be cooled below its ignition temperature before extinguishment, some feel this higher figure means that plastics are both more difficult to ignite and easier to put out. Even though some plastics produce flammable vapors, their flash points are still above the ignition temperature of wood, paper, or cotton. These authorities also point to tests that show a flame will spread somewhat more slowly on plastic surfaces than on wood, and that plastics will burn no more intensely than ordinary combustibles. This is especially true if the plastic is treated by a flame retardant or if it is covered by a nonflammable facing. Additionally, incombustible backings can markedly reduce the surface flame spread of thin

TABLE 10–4. REPRESENTATIVE MAXIMUM SERVICE TEMPERATURES FOR PLASTICS*

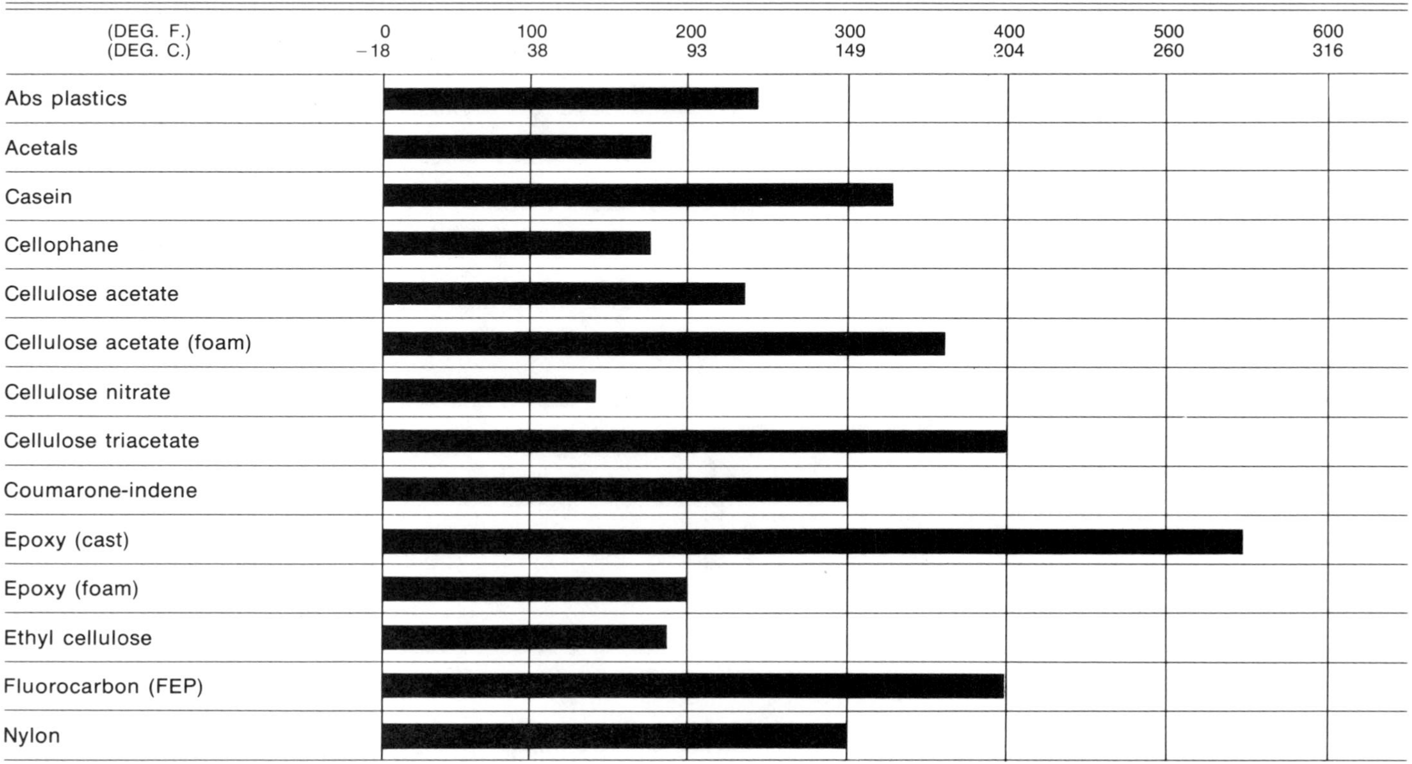

(DEG. F.) (DEG. C.)	0 −18	100 38	200 93	300 149	400 204	500 260	600 316
Abs plastics							
Acetals							
Casein							
Cellophane							
Cellulose acetate							
Cellulose acetate (foam)							
Cellulose nitrate							
Cellulose triacetate							
Coumarone-indene							
Epoxy (cast)							
Epoxy (foam)							
Ethyl cellulose							
Fluorocarbon (FEP)							
Nylon							

TABLE 10–4. REPRESENTATIVE MAXIMUM SERVICE TEMPERATURES FOR PLASTICS (CONT.)

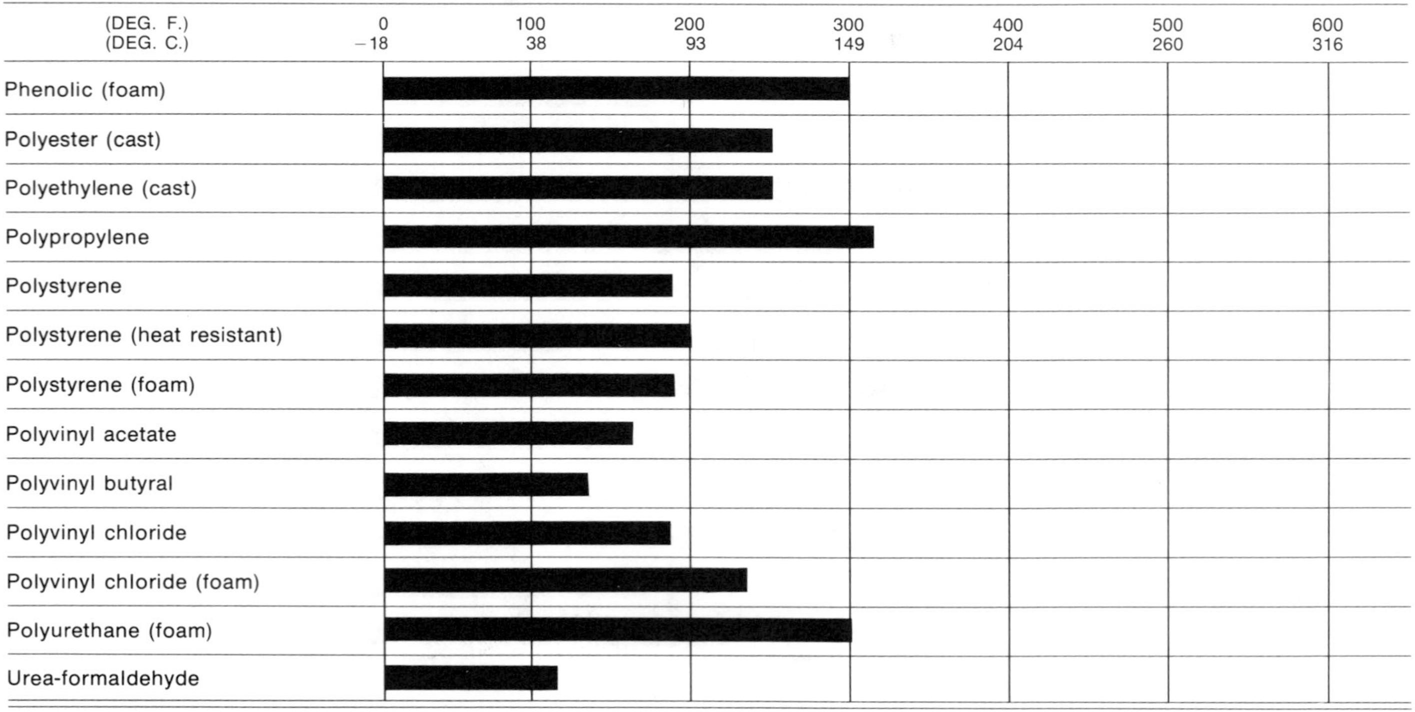

(DEG. F.) (DEG. C.)	0 −18	100 38	200 93	300 149	400 204	500 260	600 316
Phenolic (foam)							
Polyester (cast)							
Polyethylene (cast)							
Polypropylene							
Polystyrene							
Polystyrene (heat resistant)							
Polystyrene (foam)							
Polyvinyl acetate							
Polyvinyl butyral							
Polyvinyl chloride							
Polyvinyl chloride (foam)							
Polyurethane (foam)							
Urea-formaldehyde							

*These temperatures are comparative approximations and can vary widely according to the components in a particular molding compound. In each case, the top of the service range has been taken.

TABLE 10–5. THE FLASH POINT AND IGNITION TEMPERATURES OF CERTAIN PLASTICS* AND PACKING MATERIALS

	FLASH POINT		IGNITION TEMPERATURE	
	(DEG. F.)	(DEG. C.)	(DEG. F.)	(DEG. C.)
Ethyl cellulose	555	291	565	296
Nylon	790	421	795	424
Polyester	750	399	905	485
Polyethylene	645	341	660	349
Polystyrene	680	360	925	496
Polystyrene beads	565	296	915	491
Polyurethane foam	590	310	780	416
Polyvinyl chloride	735	391	850	454
Styrene-acrylonitrile	690	366	850	454
Cotton batting	490	254	490	254
Paper	445	229	445	229
Wood shavings	500	260	500	260

**Modern Plastics*, July 1961, page 119.

plastic sheets. The smooth, relatively slick surfaces of a moderately inclined plastic will also tend to shed falling embers and significantly reduce the possibility of ignition.

What tests are being cited? The burning rates of plastics have been investigated many times (see Table 10–5). In 1941, after putting cellulose nitrate in a class by itself, the Underwriters' Laboratories divided plastics into three groups.

Group 1: Those plastics which burn at a rate comparable to cellulose acetate (about like paper) and are more or less completely consumed: some acrylics, polystyrene, and cellulose acetate butyrate.

Group 2: Those plastics which burn with a feeble flame which may or may not propagate away from the source of ignition: urea-formaldehyde, casein, cast phenolformaldehyde and a few others.

Group 3: Those plastics which burn only during the application of the test flame: plasticized polyvinyl chloride, asbestos-filled molded phenolformaldehyde, a cold-molded plastic and others.

Laboratory testing has continued over the years. The results obtained by the American Society of Testing Materials are often quoted by the plastics industry itself in its literature. A small piece of plastic is clamped into position and set on fire, and measurements are taken on how it behaves. Recently, there has been growing disagreement with the conclusions drawn from these tests. Experts feel that they overly stress low flammability values, and that the phrases used—"self-extinguishing," "slow-burning," and so forth—have little bearing upon conditions experienced in a fire.

The ASTM itself says, speaking on Flammability Tests D-568, D-635 and D-1433: "These methods are intended to provide data for comparing the relative burning rates of plastics in sheet form. Correlation with flammability under use conditions is not necessarily implied."

As the AIA puts it, in their report on fire hazards of the industry, with strong support by the NFPA.

> Many of the present specifications for plastic materials refer only to flammability and thermal tests that were intended for laboratory comparison purposes as a control in the manufacture and identification of a product. Designed for small specimens, these tests often fall short in the establishment of realistic fire resistance requirements and safety restrictions for plastics in use as industrial pipelines and storage tanks or as general materials of construction. Materials listed as "slow burning" have burned vigorously in large scale "tunnel" tests which checked actual field experience, producing a rate of flame spread, fuel contributed, and smoke considerably greater than wood.

By now, most fire departments have had enough field experience with plastics in fires to add the weight of our reports on their actual behavior. A tentative agreement has already been established in some areas. Although a plastic may be "slow burning," it can also be extremely difficult to extinguish when glowing combustion burrows deep into a pile of plastic powder, granules, or beads. Some firemen have reported cases where "self-extinguishing" plastics burn freely when pre-heated by an approaching fire. There are comments about plastics which melt when ignited, forming a thick liquid that also burns. If such plastics are stored on an open mezzanine, for example, a waterfall of large flaming drops is not only impressive, but can add significantly to fire spread or be a sudden shock to any fire fighter standing on the ground floor below.

The burning rate of a plastic can be influenced by several factors. As with other flammable solids, physical form is very important. A thin film will ordinarily burn faster than a thicker piece of the same plastic. For example, tests have shown that a sheet of cellulose acetate will burn at a rate of 4 to 6 inches (10 to 15 cm) per minute, a sheet of cellulose nitrate from 10 to 25

TABLE 10–6. BURNING RATES OF PLASTIC FILMS*

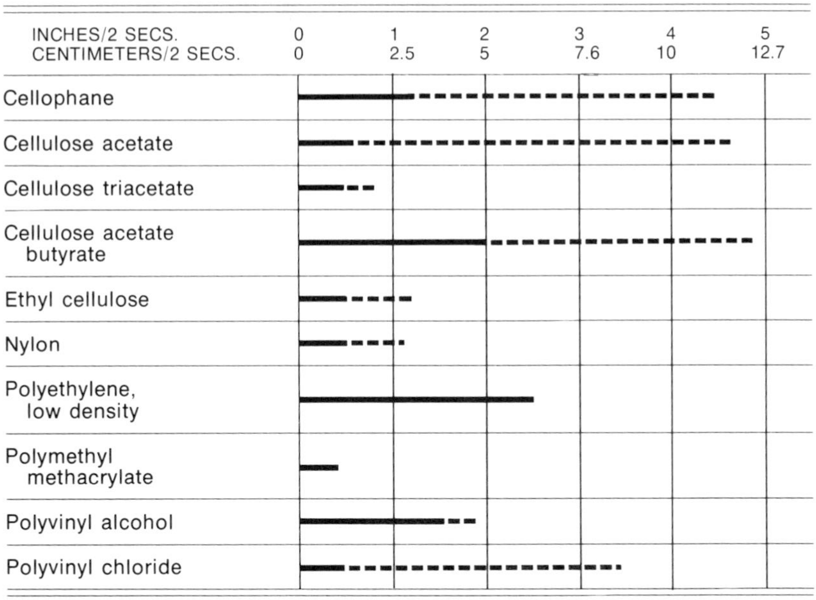

INCHES/2 SECS. / CENTIMETERS/2 SECS.	0 / 0	1 / 2.5	2 / 5	3 / 7.6	4 / 10	5 / 12.7
Cellophane						
Cellulose acetate						
Cellulose triacetate						
Cellulose acetate butyrate						
Ethyl cellulose						
Nylon						
Polyethylene, low density						
Polymethyl methacrylate						
Polyvinyl alcohol						
Polyvinyl chloride						

*Dotted portion of line represents variation in burning rate among different samples of the same materials. Tests were conducted by clamping thin film of material at a 45° angle (from horizontal) and igniting one edge.

inches (25 to 64 cm) per minute. Compare this to the burning rate of some thinner plastic films, measured in inches per 2 *seconds* shown in Table 10–6.

There is also the problem of fine division. Plastics are no different from ordinary combustible solids or metals. A plastic that is not especially flammable can become explosive when reduced to dust (Table 10–7) during fabricating or machining or if the resin is produced and handled as a powder.

TABLE 10–7. EXPLOSIVE PROPERTIES OF PLASTIC DUSTS

	IGNITION SENSITIVITY	EXPLOSION SEVERITY	IGNITION TEMP. CLOUD	IGNITION TEMP. LAYER
Acetal resins	Severe	Strong	824° F. 40°C.	
Acrylonitrile polymer	Severe	Severe	932° F. 500°C.	860° F. 460°C.
Cellulose acetate	Severe	Strong	824° F. 440°C.	664° F. 340°C.

TABLE 10–7. EXPLOSIVE PROPERTIES OF PLASTIC DUSTS (CONT.)

	IGNITION SENSITIVITY	EXPLOSION SEVERITY	IGNITION TEMP. CLOUD	IGNITION TEMP. LAYER
Cellulose triacetate	Strong	Strong	806° F. 430°C.	
Cellulose acetate butyrate	Strong	Strong	770° F. 410°C.	
Cellulose propionate	Strong	Severe	860° F. 460°C.	
Coumarone-indene resins	Severe	Severe	1022° F. 550°C.	
Epoxy resin, no additives	Severe	Severe	1004° F. 540°C.	
Ethyl cellulose, no fillers	Severe	Severe	644° F. 340°C.	666° F. 330°C.
Furane resin: phenol-furfural	Severe	Severe	986° F. 530°C.	
Melamine-formaldehyde	Moderate	Moderate	1454–1490° F. 790–810°C.	
Methyl cellulose, no fillers	Severe	Severe	680° F. 360°C.	
Nylon polymer	Severe	Strong	932° F. 500°C.	806° F. 430°C.
Phenol-formaldehyde	Severe	Strong	1076° F. 580°C.	
Polycarbonate resin	Strong	Strong	1310° F. 710°C.	
Polyethylene, high-pressure	Severe	Strong	842° F. 450°C.	716° F. 380°C.
Polyethylene, low-pressure	Severe	Severe	842°F. 450°C.	
Polyethylene terephthalate	Strong	Severe	932° F. 500°C.	
Polypropylene resins	Severe	Strong to Severe	788°F. 420°C.	
Polystyrene beads	Strong	Strong	932° F. 500°C.	878° F. 470°C.
Polyurethane resins (foam)	Severe	Strong	950–1022° F. 510–550°C.	134–824° F. 390–440°C.
Polyvinyl acetate	Moderate	Weak	1022° F. 550°C.	
Polyvinyl butyral	Severe	Moderate	734° F. 390°C.	
Rayon, viscose	Moderate	Moderate	968° F. 520°C.	482° F. 250°C.
Syrene-acrylonitrile copolymer	Strong	Weak	932° F. 500°C.	

Foamed Plastics

The surface area and consequent rate of flame spread can be greatly expanded when a plastic is foamed. Resins can be foamed in one of three ways. They can be whipped into a froth just as an egg white can be stiffened. This is called MECHANICAL foaming. CHEMICAL foaming is accomplished by having gas formed within the polymer by a reaction of some sort—an alkali reacting with an acid, for instance—or by the action of a blowing agent which forms a gas through decomposition. PHYSICAL foaming comes about through the expansion of compressed gases or volatile liquids incorporated into the resin. Many gases can foam a resin—carbon dioxide, nitrogen, various fluorocarbons, steam, air, and so forth. Ideally, the gas used should have no effect upon the flammable properties of a resin. For this reason, the use of such flammable liquids as pentane, neo-pentane, and petroleum ether in the formation of polystyrene beads has come under scrutiny. When these beads are exposed to heat, they will expand as the volatile flammable liquids in their structure turn to gas. Naturally, the hazards in fabricating polystyrene beadboard and in the storage of the raw material should be carefully considered during fire inspections and preplanning, even though the plastic may contain built-in fire retardants.

The use of foamed plastics, rigid or flexible, strong, light in weight, almost impervious to water, with excellent insulating qualities, is one of the bright horizons for the industry. Flexible foams are presently being used for cushions and the padding of furniture and ladies. Rigid foams are expected to revolutionize concepts in low-cost housing. This deserves some emphasis. Only the flame-retardant grades of foamed plastic should be used as structural materials. It is important that foamed insulations be protected from flames and physical impact damage. Foamed plastics should not be left exposed except in special situations but should be covered with a finish of some kind. When polystyrene foam melts, it shrinks away from ignition sources. Behind incombustible finishes, it will not support combustion.

Foamed plastics are very susceptible to flame spread. They have a large surface area. Polystyrene and cellulose acetate foams burn at the very respectable rate of 4.5 inches (11.4 cm) per minute. (Other types of foamed plastics include polyurethane, urea-formaldehyde, polyvinyl chloride, epoxies, and the phenolic or silicone resins which are often foamed into place.)

Foam Rubber

While we are on the subject of foamed plastics, let us not overlook another familiar material which is produced by many of the same methods: foamed latex, also called sponge or foam rubber. Synthetic rubbers, such as neoprene and butadiene-styrene, were developed during the war to replace the supplies of natural rubber cut off by the Japanese invasion of Southeast

Asia. During that time, the butadiene-styrene copolymer was known as Government Rubber-Styrene and it is still known by the initials GR-S and by such trade names as Airfoam or Foamex. Neoprene, created by the DuPont Company from a chlorinated acetylene polymer, now has several trade names. Most foam rubbers are made from GR-S or natural rubber, or from mixtures of the two.

A series of tests at the Underwriters' Laboratories showed that, at temperatures above 210°F. (99°C.) spontaneous heating of foamed natural rubber took place, accompanied by the production of flammable vapors. If the heat had been kept confined, ignition would have taken place.

These tests have caused some thought. There has been some speculation about the combination of an electric blanket and a foam rubber mattress. Every mattress fire may not be caused by a cigarette. And consider this: Some of the more secret articles which contribute to feminine allure are made of foam rubber. They have often been suspected of spontaneous heating in commercial or home dryers, leading to ignition of the whole load of clothes.

Although foam rubber burns vigorously and requires sizable amounts of water to extinguish, it burns primarily on the surface. The fire does not burrow into the interior, as it does in a cotton mattress, requiring a complete and tedious overhaul to prevent a rekindle. Quantity storage of most rubbers can lead to a fire of intense heat and rapid spread. But it is the smoke we think of first. Natural rubber is vulcanized by sulfur. Amid the dense black smoke and disagreeable odor of a rubber fire are several toxic gases.

Toxicity of Plastics

When cellulose nitrate burns or decomposes, it releases gases which can be toxic or explosive. How do other plastics behave in this regard? Like everything else, it all depends. Plastics are no different because they are man-made. Just as cellulose nitrate fumes contain the nitrogen oxides, other burning plastics send forth their own mixture of fire gases. For example, vinyl plastics create significant amounts of hydrogen chloride. The spread of this corrosive gas has been known to cause property damage far beyond the limits of the fire itself. And metallic machinery is neither as valuable nor as vulnerable as your skin and lungs, which must be adequately protected.

Plastics are all around us. Sometimes the fumes from a vinyl plastic such as polyvinyl chloride (PVC) reach out to injure or hospitalize a fire fighter completely unexpectedly, in a routine fence fire where the fence is topped with corrugated PVC. Fortunately, these fumes have an unpleasant, sharp, acrid odor which can't be missed. If concentrations build up to hazardous levels, this odor and a possible skin tingling should not be disregarded, especially since later an immediate departure from the corrosive atmosphere may be impossible.

Many other extremely toxic gases are possible in a plastic fire: ammonia, some of the cyanides, phenol, aldehydes, amines. That they are generally produced in small amounts has led to some conclusions stated in Underwriters' Laboratories Research Report 53.

> It is recognized that the combustion and thermal decomposition of plastics materials may present a hazard to life under fire conditions, but in the opinion of these authors the chief hazards in this connection are due to the presence of carbon monoxide or atmospheres lacking in oxygen, which is likewise the case where wooden or other cellulose materials are involved in a fire.

Do not, even briefly, consider this as minimizing the dangers of the fire gases which plastics can produce. Rather, it is another recognition of the properties of carbon monoxide, the major cause of fire death in the United States. To a fire fighter, anything smoking may be hazardous to his health, including plastics.

Some plastics are not far behind rubber in the amount of smoke they will produce. Anyone who has burned a test piece of foamed polystyrene or witnessed a fire in a sizable amount of it will testify to this. The presence of carbon monoxide can even be predicted from the smoke—black—a color created by tiny particles and clumps of unburned carbon, another signal of incomplete combustion. Some plastic fires create a dense, blinding pall which makes any entry a challenge. Many a fire report emphasizes the presence of an astonishing amount of smoke.

> [In an $800,000 fire in Minnesota,] the fire quickly spread through the area. The burning plastics generated huge amounts of black smoke as the fire grew. Firemen could not find their way into the building to get to the seat of the fire, because of the obscuring, dense smoke.

> [In a polystyrene fire in New York,] the fire spread with such rapidity that hundreds of sprinkler heads were opened and the water supply was insufficient to extinguish the fire. The fire department encountered large quantities of dense black smoke which hampered fire fighting. Flames shot up to a height of 50 feet.

> [At a polystyrene foam fire,] the entire interior of the building was involved when the firemen arrived. The contents were subject to a heavy smoke damage.

> [At a polyurethane foam fire,] due to the large amount of dense smoke, it was impossible for the fire department to locate or determine the extent of the fire. Sprinklers were, therefore, left on for about an

hour and were turned off only after a 20′ x 20′ hole was cut in the roof to release the smoke.

Identifying Plastics

Even with plastics, we cannot escape the need to identify materials. Polystyrene does not react like polyvinyl chloride in a fire, and neither resembles cellulose nitrate. Without identification, the potential of a fire is unknown. Instead of developing a fire plan that takes the cause and the possible hazards of the materials involved into consideration, we find ourselves responding to the effects as they happen. The fire remains a jump ahead of us, controlling our reactions, instead of the opposite. The positive identification of a particular plastic is difficult. Although we will briefly identify about twenty groups of plastics in the pages immediately following and include some of their uses and trade names, there are reasons why this will be of limited help.

1. The field grows faster than books can be written; not only are new plastics constantly introduced, but so are improvements upon old varieties.

2. Several different plastics, all looking much alike, can be used for the same purpose. Turning it around, the same plastic can be produced in many forms, from films to shapes to foams, with its properties altered by its structure.

3. Trade names may not be too helpful either. A familiar brand name may only reflect the size of the advertising budget of the parent company, not the amount in use in your particular area. As we saw in the case of cellulose nitrate, many trade names can refer to the same material. It is also possible for the same trade name to refer to completely different types of plastic.

With all these limitations in mind, let us see what plastics are being manufactured in quantity these days and how we can identify them.

ABS Plastics

Developed in 1948, these thermoplastics combine rigidity with high impact strength. They are made from the copolymerization of three monomers: acrylonitrile (a poisonous, unstable, flammable liquid), butadiene (an asphyxiating, unstable, flammable gas), and styrene (an irritating, unstable, flammable liquid). ABS plastics are available for subsequent processing in the form of powder, granules, or sheets. Trade names: Lustrex, Royalite, or Uscalite. Typical uses: pipe and fittings, wheels, football and safety helmets, refrigerator parts, battery cases, tote boxes, water pump impellers, utensil and tool handles, radio cases. A moderate fire hazard, ABS plastics are slow burning, do not drip, and have an odor of illuminating gas.

Acetal Resin

Acetal resin is a rigid thermoplastic developed in 1956 for use in fields now dominated by diecast metals. It is extremely tough, has outstanding tensile strength, stiffness, and fatigue life, retaining these properties under adverse conditons of temperature and humidity. It is made from the polymerization of formaldehyde (a toxic, unstable, flammable gas). Acetal resin is produced in powder form, under the trade name of Celcon, a copolymer, and Delrin. Typical uses: automobile instrument clusters, carburetor parts, gears, bearings and bushings, door handles, plumbing fixtures, and moving parts in home appliances and business machines. It burns with a clear blue flame, melting and producing flaming drops.

Acrylics

Acrylics are a group of thermoplastics which combine strength and rigidity with exceptional clarity and good light transmission. In crystal-clear form, acrylic plastics can "conduct" light, pick it up at one edge and transmit it unseen, even around curves. The various acrylics are made by the polymerization of acrylic acid (a poisonous, unstable, flammable liquid), methacrylic acid (a corrosive, unstable, combustible liquid), acryonitrile, or other chemically related compounds. Acrylics are available as rigid sheets, rods, tubes, and molding powders for eventual use as airplane canopies and windows, TV and camera viewing lenses, dentures, brush backs and combs, costume jewelry, lamp bases, scale models, outdoor signs and counter displays, automotive tail lights, skylights and paints. The Houston Astrodome is covered by a double layer of acrylic panels. Plastic products can be made several ways, including the forcing of the polymer through the holes of a spinnaret to form a fiber. Trade names for the resin include Lucite and Plexiglas. The fibers have such household names as Acrilan, Dynel, Orlon, and Verel. Acrylics ignite readily, soften, but generally do not drip when burning. The flame is blue with a yellowish-white tip. The smoke may have a sweet or fruity odor.

Amino Plastics

The two major types of these thermosetting plastics are formed by the reaction of formaldehyde with either melamine (a colorless crystalline solid which may release toxic gases when heated) or urea (a white crystal formed industrially by combining ammonia with carbon dioxide, or internally by our urinary system). Together, these plastics have reached an annual production level of over 400 million pounds (181 million kg). They are color-fast, glossy, scratch-resistant, very hard, and unaffected by many common solvents. Amino plastics are available as foams, white molding powders or granules with such trade names as Cymel or Bettle, or as clear, syrupy resins. Molded products are formed in the presence of acid under the influence of heat and pressure. Typical uses for melamine-formaldehyde: tableware (Melmac), distributor heads, buttons, baked enamel finishes, textile and paper treating. For urea-formaldehyde: lamp reflectors, radio cabinets, plywood adhesives (Griptite and Weldwood). Maximum service temperatures for melamine range between 210°F. and 400°F. (99°C. and 204°C.), depending upon the filler. For urea-formaldehyde, 170°F. (77°C.). The amino plastics will swell and crack when exposed to a flame and produce a smoke which smells like formaldehyde, ammonia, urea, and fish, but they are considered to have a low flammable hazard.

Casein

Many naturally occurring substances such as shellac, amber, pitch, or asphalt are plastics. Plastics can be made from protein, whether the source is animal (hides, bones, hair, horns, or hooves), or vegetable (wheat, peanuts, soybeans, or corn). Casein is the protein found in milk. When it reacts with formaldehyde, it forms a thermosetting plastic (Ameroid) which is available in sheets, rods, and tubes, or as a powder or liquid. Casein was one of the first plastics to be introduced—in 1919. Although strong and rigid, it does not withstand moisture or temperature changes too well. It ignites, swells, and chars when contacted by a flame, producing an odor of burnt milk—logical if unpleasant.

Cellulose Plastics

Cellulose nitrate, formed by the action of nitric acid upon cellulose, was the first acceptable man-made plastic. Since its creation in 1868, many other chemicals have been put to work upon cellulose. By now, they have formed a large group of usable thermoplastics. Cellulosics are among the toughest of plastics, retaining their lustrous finish under rough usage. But with a few exceptions, they weather poorly, their maximum service temperatures are rather low, and they must be kept away from alkalis and alcohols. Despite their limitations, the cellulose plastics are available in most of the common forms for a wide variety of uses.

Cellulose Acetate (C/A) Made by treating cotton linters or wood with acetic acid or acetic anhydride, both of which are highly irritating and flammable liquids, C/A is used today for films (DuPont, Kodapak I), such finished articles as washable playing cards and toys (Plastocele), as a textile fiber, or as an expanded foam. Cellulose acetate burns at about the same rate as paper, produces sooty black smoke, a vinegar odor, and a dark yellow flame; it softens and forms bubbles and drops that burn as they fall.

Cellulose Acetate Butyrate (CAB) Made from cellulose after treatment by butyric acid (a poisonous and flammable liquid), acetic acid, and similar anhydrides. One of the more weather-resistant of the cellulosics, it is used in greenhouses and outdoor signs. Trade names include Fairylite and Kodapak II. CAB is considered to be a moderate fire hazard, igniting readily, melting and producing burning drops. The flame is blue with a yellow tip and sparks. Black smoke is created, along with a rancid butter odor.

Cellulose Propionate (CP) Made from cellulose reacting with propionic acid (a highly irritating and flammable liquid) and acetic anhydride, CP is used for housings for hand telephones, pens, and pencils, and has such trade names as Forticel and Tenite. It burns with a blue flame, a yellowish tip, and sparks; melts, drips fire; and creates a fragrant aroma.

Cellulose Triacetate (CT) Burns at a rate which varies with the plasticizer used in manufacture. It will melt, turn black, and produce an acrid smell. But the flammability rate is quite low and CT has a relatively high maximum service temperature of 400°F. (204°C.). It is used in recording tapes, safety glasses, visual aid transparencies, as a textile fiber, and in Safety Base film. Kodapak IV is such a film.

Film made of cellulose acetate burns about like cellophane, a little less than 2.5 inches (6 cm) per second. Cellulose triacetate film burns more slowly, at a rate less than 0.5 inch (1 cm) per second. Compared to the fierce burning of cellulose nitrate film, either acetate burns peacefully. They are far safer in other ways: ignition temperatures of the acetates are 500 Fahrenheit degrees or 278 Celsius degrees higher than nitrate; the decomposition temperature is higher; decomposition fumes are much less toxic, and, very important, decomposition only proceeds when there is a continued source of heat.

Since 1952, all photographic film in this country has been made from Safety Base, cellulose acetate or cellulose triacetate. However, nitrate film is still around. It is still used abroad, for example; in many movies being shown today in art houses or revivals; or in film on the late-late show dating from before 1952. Converting this nitrate film to Safety Base is not considered worth the cost. Consequently, the problems of nitrate film storage are not over yet.

Recommended safety measures include the segregated storage of nitrate film in separate vaults. Vault doors, reel bands, reel cans, and storage records should be clearly identified with the words "Nitrate Film" in red lettering, "Safety Film" in bright green. When in doubt about a particular reel of film, take a small piece outside and burn it. Doubt will not last long. The words "safety base" are often printed along the edges of the film strip itself to help identify it.

Ethyl Cellulose (E/C) E/C is a molding compound created when a form of cellulose is exposed to an ethylating agent like ethyl chloride (an anesthetic and flammable liquid or gas). It is used in electrical parts, flashlights, and hose nozzles. Igniting readily, it burns with a greenish-yellow flame, has drips which also burn, and creates an odor of burnt sugar. A similar plastic is methyl cellulose, which is also flammable. This plastic can be used to package a powder until the package and the powder dissolve together in water. Trade names: Ethocel or Methocel.

Regenerated Cellulose Made when a form of wood is treated by sodium hydroxide (a nonflammable, but corrosive liquid or solid), carbon disulfide, and an acid. The familiar brand name for this material is Cellophane, and the uses are equally familiar: bags, pressure sensitive tape, wrappings, etc. Cellophane burns rapidly, with an odor of burning paper.

Plastic Textiles

When various forms of cellulose are dissolved, processed, and forced through a spinnaret to form a thread, they can be woven into a familiar fabric: RAYON. Trade names for the various rayons include Arnel (triacetate rayon), Chromspun (acetate rayon), Matesa (cuprammonium rayon), Cordura (viscose rayon), and Fortisan (for rayons spun from cellophane). Plastic fibers are no different from natural fibers. If they are flammable, they will burn at a rate that is partially determined by their weave and length of nap. The so-called "torch sweaters" of two decades ago were made of brushed rayon. Once ignited, these sweaters were completely consumed in 20 to 40 seconds. The shaggy cowboy pants, so beloved by small boys, have been the direct cause of several deaths when they ignited and could not be removed in time. AIA Interest Bulletin 264, has this to say:

> Only a small proportion of textiles are so flammable as to be dangerous when worn by persons. These are mostly the long-piled, brushed, or napped fabrics, certain very sheer fabrics, and nets which are stiffened with a solution of cellulose nitrate, and some unsupported plastic films and coated fabrics which are used for aprons and raincoats. Although many articles of clothing ordinarily worn are combustible and may burn readily when ignited by flame or other source of heat, they do not flame to the extent that the wearer cannot extinguish the flame quickly or is prevented from quickly removing

> the garment. [When a material] is prepared in the form of fine, loose, and fleecy fibers, it is surrounded by a liberal supply of air. It will ignite quickly and be highly flammable. This extreme ease of ignition and instantaneous flame spread over fabric surfaces present a most hazardous situation. The panicky wearer under such circumstances cannot extinguish the flames nor promptly divest himself of the flaming garments.

What is true of clothing is also true of some plastic fibers, such as the acrylics, which are woven into rugs. Rugs with a long fluffy pile have been known to greatly assist the spread of fire in multiple-occupancy dwellings, carrying the flames down hallways, up stairways, and beneath doors to uninvolved areas.

Cold Molded

Cold molded plastics are thermosetting materials produced when an inorganic binder (cement, glass, lime, ceramics) or an organic binder (phenolic resin, bitumin) is mixed with a filler, usually asbestos. The mixture may then be placed in a cold mold under pressure and subsequently baked. Cold molded plastics with organic binders are very difficult to ignite; those with inorganic binders will not burn. Some of these cold molded plastics have very high continuous service temperatures, up to 1500°F. (816°C.). Brand names: Tico, Garit, Hemit, or Robit. When heated, they may emit a waxy odor, but there is no sign of melting or dripping.

Coumarone-Indene Resins

One boast of the industry is that these man-made materials are derived from only six basic sources: air, water, salt, cellulose, petroleum, and coal. Coumarone-indene resins are obtained by heating the light-oil fraction of coal tar. The two liquids are then polymerized with the aid of sulfuric acid, although indene merely needs the help of air and sunlight. Coumarone-indene thermoplastics are used in combination with rubbers, other synthetic and natural resins, or waxes. Trade names include Cumar, Loxite, Neville, and Picco.

Epoxies

First manufactured in 1947, the epoxies are a series of thermosetting resins produced by the copolymerization of epichlorhydrin (a toxic, unstable, flammable liquid) with a phenol or glycol. Epoxies are available commercially as molding compounds, resins, foamed blocks, or liquid solutions bearing such trade names as Cardolite, Durafoam, Fiberlite, and Hysol. With such properties as flexibility, adhesiveness, and chemical resistance, most of the annual production goes into paints, varnishes, and glues. Epoxies ignite readily, charring as they burn.

Fluorocarbons

A cylinder of tetrafluorethylene gas (TFE) unexpectedly polymerized spontaneously into a white powder one evening in 1943. The chemists gathered 'round. TFE is built about like ethylene, except that fluorine atoms have taken the place of the hydrogen atoms. The gas had formed a solid polymer similar to polyethylene. (Compare Figure 10–7 below with Figures 10–2 to 10–4.)

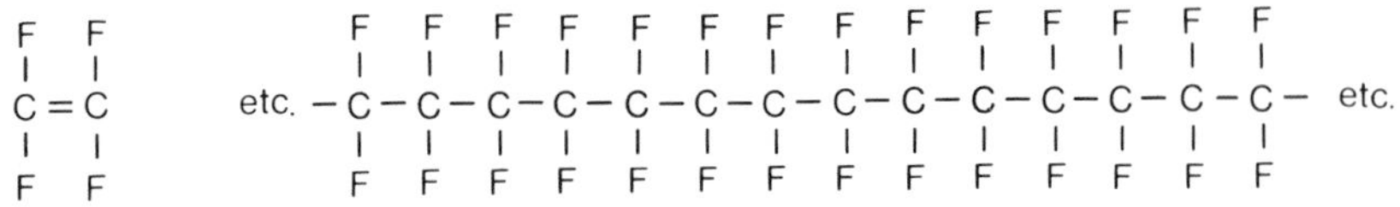

```
F   F            F   F   F   F   F   F   F   F   F   F   F   F   F   F
|   |            |   |   |   |   |   |   |   |   |   |   |   |   |   |
C = C     etc. - C - C - C - C - C - C - C - C - C - C - C - C - C - C -  etc.
|   |            |   |   |   |   |   |   |   |   |   |   |   |   |   |
F   F            F   F   F   F   F   F   F   F   F   F   F   F   F   F
```

Tetrafluoroethylene
C_2F_4

Polytetrafluoroethylene

FIGURE 10–7. Teflon

This new plastic, Teflon, had some remarkable properties. It was strong, extremely hard, had a high impact strength and, for a plastic, was very resistant to temperature extremes. Most fascinating, it was so slippery that practically nothing would stick to it. Teflon could be used to coat frying pans, electric irons, ovens, and other household appliances. Liquids rolled right off this unique plastic that resisted both oil and water. It became possible to create a fluorocarbon with a sticky side and a slippery side. This coating (Zepel, Scotchgard) could be used on fabrics to make them water- and stain-repellent. Several other fluorocarbon polymers and copolymers have emerged over the years. They are used in such industrial applications as valve seats, gaskets, pump diaphragms, linings, tubing, and high voltage insulation.

Fluorocarbon plastics *will not burn*, but they can melt, char, bubble, and decompose when exposed to high amounts of heat. The gases formed during decomposition are very poisonous. Wear a self-contained mask if fluorocarbon plastics are involved in a fire.

Nylon

Nylon is now the chemical name, the generic name, for a related group of thermoplastic resins also called the polyamides or called by such trade

```
      O  H  H  H  H  O                       H  H  H  H  H  H  H  H
      ‖  |  |  |  |  ‖                       |  |  |  |  |  |  |  |
etc.- C -C -C -C -C -C -O-H   +   H-N-C -C -C -C -C -C -N-etc.
         |  |  |  |                             |  |  |  |  |  |
         H  H  H  H                             H  H  H  H  H  H
```

Adipic acid **Hexamethyldiamine**

```
                   O  H  H  H  H  O  H  H  H  H  H  H  H  H
                   ‖  |  |  |  |  ‖  |  |  |  |  |  |  |  |
H-O-H +   etc.- C -C -C -C -C -C -N -C -C -C -C -C -C -N-etc.
                      |  |  |  |        |  |  |  |  |  |
                      H  H  H  H        H  H  H  H  H  H
```

Water **Nylon**

FIGURE 10–8. Formation of Nylon

names as Tynex and Zytel. It is formed by the condensation polymerization of certain acids with diamines. For example, adepic acid and hexamethyldiamine will condense together after forming a water molecule. This is easier done than said.

Nylon is a very versatile plastic. It is not only used in hosiery and clothing, but also in fishing lines, brush bristles, rugs, zippers, washers, gears. Nylon is difficult to ignite. More frequently it will melt, drip, froth, and produce an odor variously described as celery, wool, or burning leaves.

Phenolics

There is a little bug, a native of Southeast Asia, that at one stage in its life retires behind a resin which it creates from the sap of a tree. Over 150,000 of these lac bugs must be harvested to produce a single pound of shellac. At the beginning of this century, Dr. Leo Baekeland began looking for a synthetic replacement for shellac. He began experimenting with a combination of phenol and formaldehyde. Over 40 years had passed since the introduction of celluloid, and cellulose nitrate was merely the rearrangement of an existing molecule. What Dr. Baekeland eventually created was a brand new molecule, entirely man-made, which had no duplicate in nature. He did not stop with the discovery. Dr. Baekeland developed techniques for putting his phenol-formaldehyde resin to work as a phenolic which could be cast—one that could be formed under heat and pressure—and a solution which could be used in laminates. Small wonder the first trade name for this new material honored him: Bakelite. Today, additional brand names include: Formica, Fiberfil, Resinox. The trade name "Bakelite" now refers to many different groups of plastics.

Phenolics are tough and strong, the real "work horses" of the plastics industry. Typical uses are distributor heads, telephones, radio and TV cabinets, insulation, home appliances, drawers, dials, handles, knobs, brake linings, bondings, coatings. Several types of phenolics are produced from the reaction of the various phenols and aldehydes. Different fillers, such as mica and asbestos, will produce compounds which neither burn, melt, nor drip. *Phenolic foams* are considered to be self-extinguishing. Thermosetting cast phenolic will ignite, but it burns slowly. It gives off the odor of formaldehyde and phenol. Phenol, also called carbolic acid, is a poisonous flammable solid or liquid.

Polycarbonate Resin

Discovered (also accidentally) in 1957, this transparent thermoplastic (Lexan or Merlon) is the toughest of all plastics. A sledge-hammer blow will not shatter it. Thin sheets of Lexan will stop bullets. This polycarbonate has already found such interesting uses as space helmet visors, unbreakable windows, and aircraft gauges. The resin is considered to have a low flammability rating.

Polyesters

Polyesters were first developed in 1942. They are all thermosets with one exception, familiarly known as Mylar recording tape. These resins can be used to impregnate cloth and other materials, to reinforce plastics, or to create synthetic fibers (Dacron and Kodel). They can be cast or molded into automobile bodies, luggage, skylights, translucent roofs, and decorative interior partitions. Cast polyester will ignite, melt at the edges, and produce quantities of black smoke and a faint, sweet odor. Maximum service temperatures range between 250°F. and 450°F. (121°C. and 232°C.), depending upon the filler used.

Polyethylene

At last count, a single plastic, polyethylene, made up almost one-third of the industry's total production. The number of uses to which this waxy, lightweight thermoplastic is put would fill many pages. Here are only a few: food and textile packaging, rigid and squeezable bottles, acid-resistant tank linings, bristles, bowls, garbage cans, wastebaskets, toys, electrical insulation, pipe, coatings on paper and other materials, flexible ice cube trays, glasses, dishes, carboys, rain capes, meteorological balloons, greenhouses, silo covers, cable jacketing, moisture barriers under concrete, walls, cordage, filter cloths, upholstery, and artificial flowers and grass. Trade names include Polyfilm, Spunglow, Visqueen, and Surlyn, the latter a new see-through type. One reason for the popularity of polyethylene is the way it can be altered to fit a particular need. The polyethylene chain can be made to branch or cross-link. When this happens, the plastic will change its density, flexibility, and resistance to heat. But even at its best, it should be kept out of ovens. The maximum service temperatures range between 165°F. and 300°F. (74°C. and 149°C.). Polyethylene melts as it burns, and the drippings continue to burn. Once ignited, it burns rapidly with a blue flame tipped with yellow, giving off an odor like burning wax.

Polypropylene

First introduced into the United States in 1957, polypropylene is formed, as we might expect, by the polymerization of propylene (an anesthetic, flammable gas). It has better heat resistance than polyethylene and can be used in sterilizable bottles and containers, for molded products (Moplen) which have an unusual chemical resistance, in fibers (Firestone), and in transparent films (Propylene) that are impermeable to vapors and gases. This thermoplastic ignites, burns slowly, melts and drips, and produces an odor not unlike heated asphalt.

Polystyrene

I. G. Farben* chemists first made polystyrene in 1929. It can be processed into rigid forms (Styron, Pliolite, Cerex), into films (Tricite, Styrofilm), or into foams (Styrofoam, Scotbord). Molded polystyrene is widely used in containers, battery cases, radios and the like; the foam in toys, insulation, counter displays, and package cushioning; or foamed-in-place. The maximum service temperatures for polystyrene are rather low. Even the so-called heat-and-chemical-resistant types do not withstand continuous exposure to temperatures much above 200°F. (93°C.) Polystyrene ignites easily, softens, bubbles, and burns with the formation of impressive amounts of black smoke. Burning polystyrene has an odor variously described as that of illuminating gas or marigolds. Polystyrene is a thermoplastic.

Polyurethane

During World War II, the Luftwaffe needed strong wing tips, and with the way the Spitfires were acting, they needed them at once. German technology provided a short-term assist by developing polyurethane foam. In recent

*I. G. Farben is not a person's name. It stands for the German "Industrie Gesellschaft Farben," Industrial Dye Corporation. This was a pre–World War II industrial combine or cartel which produced everything from photographic film (AGFA) to medicinal products.

years, industrial interest in these plastics has expanded rapidly, as the number of trade names testifies: Curifoam, Eccofoam, Genfoam, Pliofoam, Polyfoam, Polylite, Polyrubber, Scotchfoam, Selectrofoam, Stafoam, Vibrafoam, and many others. Flexible urethane foams have been used for some time in cushions, mattresses, rug underlays, sponges, mats, and so on. Rigid urethane foams are now coming into their own as highly efficient insulators. They are the reason why the walls of our refrigerators are thinner than they were formerly. In addition, they can add strength to structures without a great deal of weight, are highly buoyant, and can be foamed in place. Urethane foam reinforced by a polyester (Corfam) can even substitute for leather. Or, take a good look at your cigarette filter.

One of the leading manufacturers of the raw materials, Union Carbide Company, has several interesting comments about rigid polyurethane foam.

> Rigid urethane foam is formed by the reaction of two liquids in the presence of a gas-producing blowing agent, usually carbon dioxide (or a fluorocarbon). As the chemical reaction occurs, heat is generated and the blowing agent vaporizes to form tiny bubbles in the thickening plastic. In less than 2 minutes, the foam expands to its full height, and sets. What is essentially (formed) is one giant cross-linked molecule containing entrapped bubbles of gas.
>
> Rigid urethane foam has inherent adhesion properties as it foams. This adhesion is so tenacious that rigid urethane foam can be applied by spraying. A foam-producing operation should be well ventilated. Full employee protective equipment is needed and it is frequently wise to use chemical cartridge respirators. Cured rigid urethane foams are inert and do not represent a toxicological hazard.
>
> Polyruethane foams are combustible; however, rigid urethane foams can be made so that they will not support combustion, or will not burn through when heated with a propane torch. Such resistance to burning is achieved at either increased cost or sacrifice in other properties. It is always desirable to test rigid urethane foam under the actual condition to be encountered in use. Like many plastics, urethane foam has a softening point. This softening point appears to be near 250°F. (121°C.).

Polyvinyls

The thermoplastic group of plastics known as the polyvinyls or vinyls are another member of the billion-pound-a-year club, ranking second in amount of use only to polyethylene. They are produced in several forms: flexible, rigid, as fibers, or as foams. All types of vinyls are tough and strong. Although they will stand up to household temperatures, they should be kept away from any contact with heat sources. Most of them are slow burning; a few will not sustain combustion unless in contact with a flame. Table 10–8

TABLE 10–8. VINYL PLASTICS

	MAXIMUM SERVICE TEMP.	TYPICAL USES AND TRADE NAMES	FIRE CHARACTERISTICS
Polyvinyl acetal	125–150° F. 52–66°C.	Adhesives, inks, plastic wood. Rigid: INSULAR, LEMAC	Low flammability: ignites, softens, yellow flame sooty smoke, vinegar odor. The vinyl acetate monomer is an unstable flammable liquid, with low toxicity.
Polyvinyl alcohol		Water-soluble packages for dyes, soaps, detergents. Rigid: ELVANOL, RESISTOFLEX	Moderate flammability: ignites readily, softens, melts, spatters, blisters. Sweet floral odor.
Polyvinyl butyral	100–140° F. 38–60°C.	Inner layer in safety glass. Rigid: BUTACITE, SAFLEX	Low flammability: ignites, softens, melts, drips; blue flame with a yellow tip; rancid butter odor. The vinyl butyral monomer is a fairly toxic flammable liquid.
Polyvinyl chloride (PVC)	120–175° F. 49–79°C.	Automobile seat covers, flooring, upholstery, wall and floor coverings, corrugated fence toppings, tarpaulins, packaging. Rigid: VELON, RESINITE Films: NAUGAHYDE, PRESTOFLEX, KOROSEAL Foams: U.S., VINACEL, VINYLFOAM	Low flammability: burns in contact with a flame. Softens under heat and produces a toxic white smoke and a chlorine odor. When encountered (see list of uses) respiratory protection must be worn. The vinyl chloride monomer is an anesthetic, unstable gas which is frequently found liquefied.
Polyvinyl chloride-Polyvinyl acetate copolymer	150–175° F. 66–79°C.	Textile fibers, baby pants, shoes, rainwear, shower curtains, phonograph records, coatings. Fiber: VINYON Films: VINYLITE, FABTEX, BAKELITE	Low flammability: difficult to ignite, softens, chars, bubbles; smoky yellow flame; noticeable odor of acrid hydrochloric. Respiratory protection must be worn.
Polyvinylidine chloride	160–200° F. 71–93°C.	Upholstery fabric, screening, draperies, food wrap. Saran Wrap is a copolymer with PVC. Rigid: GEON Fiber: SARAN, VELON	Low flammability: difficult to ignite, softens, chars, odor of chlorine. Respiratory protection must be worn. The vinylidine chloride monomer is a moderately toxic, unstable, flammable liquid.

lists some of the uses, trade names, and flammability characteristics of the common vinyl plastics.

Silicones

Developed in 1943, the silicones—so called because their structure is based upon silicon rather than carbon—are available as resins, coatings, greases, fluids, and silicone rubbers. Their degree of flammability is determined by their filler—either slow burning or completely nonburning. Up to now, the big outlets for the silicones have been adhesives, cosmetic ingredients, and such electrical parts as switches, insulation, coils, and power cables. Typically, research has uncovered some exciting new uses for them. Silicones are significantly more resistant to heat than other plastics. New types have been produced which withstand the tremendous frictional heat of rocket re-entries by slowly decomposing, layer by layer. The process is called ablation. Silicone rubbers in extremely thin sheets will allow the passage of oxygen, but not water. Some day, undersea cities may be enclosed in these plastic films, or perhaps they can be used to replace lung tissue. Like the present, the future of mankind, wherever we go, may well be packaged in plastics.

Summary

The brave new worlds of plastic are not our real concern. What confronts us is sorting out and identifying individuals and groups amid the avalanche of polymers so that we can make some sort of an intelligent fireground decision. Just as a golf course is the worst place to learn how to play the game, a fire scene is not the time to suddenly learn a plastic produces toxic smoke or that its burning rate is beyond expectation, or even reason. We must find out what *may* happen *before* it does! This is why we have considered the major types of plastics, trying to discover how and where they are used, what trade names they bear, and how they react in a fire. Preplanning begins with identification.

Although much is still unknown about the behavior of plastics in a fire situation, we do know that many of them are flammable. Their rate of burning can be changed by their form, and our tactics must take this into account. We know that toxic hazards exist: carbon monoxide primarily, but also other unfriendly gases and the possibility of dense concentrations of smoke. These are considerations that demand a place in our sizeup. We also know that certain resins, such as the epoxies and the isocyanates (polyurethane), become more toxic when heated. A good rule of thumb is to suspect all plastics of plotting an assault upon our respiratory systems until they are proven innocent. We should protect ourselves at all times with at least a self-contained gas mask. (Incidentally, if a corrosive gas is released by a plastic such as one of the chlorinated vinyls, warn the owners that metallic equipment may have to be promptly cleaned or neutralized. Their insurance company will be grateful.)

With the exception of cellulose nitrate, shipping and storage regulations reflect the conclusion that plastics have a degree of hazard similar to ordinary Class "A" materials. Why, occasionally, do they cause such trouble? All too often, the answer is mishandling. The industry itself has stressed this fact for years. A plastic is designed with a set of properties which *must* be respected. Time and again plastics are misused in ways which their characteristics make dangerous. Consider the following examples:

1. A warehouse owner allows great heaps of polystyrene beads to lie about in open bins separated from an ignition source only by good fortune.

2. A builder doesn't use fire retardant grades of foamed insulation or leaves them exposed to physical damage and flame. Runs of flammable plastics are not interrupted by the separation needed to reduce an area to hand-line size.

3. A housewife puts out a glassy-looking thermoplastic coaster and a guest mistakes it for an ashtray. Her husband, the home handyman, installs thermoplastic wall tiles around his stove. He is so proud of their decorator colors he never notices them beginning to curl toward the oven.

And don't think fire fighters are exempt. We know that water is recommended for a plastic fire. But it must be used wisely. We are not above directing a straight stream into a pile of plastic dust and powder, causing the formation of an explosive cloud. Also, a plastic fire is very likely to burrow into a pile, awaiting a time to rekindle. The premature closing of sprinkler systems and hose lines has been a source of regret for many fire officers.

QUESTIONS ON CHAPTER 10

1. What are two names often given to cellulose nitrate plastics?
2. Discuss the rate of combustion of cellulose nitrate.
3. What are some visible signs that decomposition of cellulose nitrate is taking place?
4. Name three gases produced by decomposing cellulose nitrate. How many are flammable?
5. What should be done if cellulose nitrate is spilled inside a truck in transit?

6. Name three ways cellulose nitrate can cause an explosion.
7. What is the recommended extinguishing agent for cellulose nitrate?
8. What is a monomer? A polymer?
9. What is the difference between a thermoplastic and a thermoset?
10. Why is the maximum service temperature of a plastic important to a fire fighter?
11. Name a plastic which can melt in a fire and produce flaming drops.
12. Name a plastic which produces great quantities of black smoke when on fire.
13. Name a plastic which can produce a corrosive gas when heated or on fire.
14. What identification measures are recommended for cellulose nitrate film?
15. Discuss some of the hazards of the spontaneous heating of foam rubber.

BIBLIOGRAPHY

"Giant Molecules," Time-Life, Inc. (1967). An extremely interesting general survey of the plastics industry.

Basic source for information about plastics: *Modern Plastics Encyclopedia* (1967).

" 'Plastic Fires' Create New Hazards for Both Firemen and Public," *Journal of the American Medical Association* (December 22, 1975).

"Sensory Irritation Evoked by Plastic Decomposition Products," *American Industrial Hygiene Association Journal* (October 1974).

"Safety and Plastic Conduit," *New England Electrical News* (May 1974).

Union Carbide Company:

"Rigid Urethane Foam" (1965).

"Bakelite Phenoxy Resins" (1965).

DuPont Company:

"Nitrocellulose" (1965).

Dow Chemical Company:

"Plastics as Materials of Construction," an address given by R. W. Theobald (1966).

Factory Mutual System:

Chapter 63, "Plastics," *Handbook of Industrial Loss Prevention,* 2nd Edition (New York: McGraw-Hill Book Company, 1967).

National Fire Protection Association:

Standard No. 40, "Storage and Handling of Cellulose Nitrate Motion Picture Film."

Standard No. 42, "Storage, Handling, and Use of Pyroxylin Plastic in Factories."

Standard No. 43, "Storage and Sale of Pyroxylin Plastic in Warehouses, Wholesale, Jobbing, and Retail Stores."

Standard No. 231, "Recommended Safe Practices for General Storage."

Standard No. 654, "Prevention of Dust Explosions in the Plastics Industry."

Handbook: Section 6, Chapter V; Section 7, Chapter VI; Section 8, Chapter VIII.

American Insurance Association:

Research Report No. 1, 1963, "Fire Hazards of the Plastics Manufacturing and Fabricating Industries."

Special Interest Bulletins:

No. 1, July 1933, "A Costly Lesson in Fire Prevention."

No. 49, December 15, 1937, "Nitrocellulose Fires."

No. 173, August 2, 1943, "Occupancy Hazards."

No. 264, February 15, 1952, "Fire Hazards of Flammable Fabrics."

No. 281, September 16, 1949, "Fire Problems of Plastic Novelties and Ornaments."

No. 283, October 25, 1950, "Film, Motion Picture, Cellulose Acetate."

No. 287, May 31, 1951, "Foam Rubber."

VISUAL AIDS

"Synthetic Fibers," 15 minutes, black and white, and "Plastics from Petroleum," 12 minutes, color, available from many college film libraries, are two films for classes interested in the production end of the industry.

DEMONSTRATIONS

With the cooperation of a local industry, you can present to your class two very interesting demonstrations. The burning rate of cellulose nitrate is an eye-opener. Unless you use very small amounts, this demonstration may have to take place outside. Its combustion rate and the way a fire will spread is clearly demonstrated if you make a short trail of the cottony material.

Polyurethane foaming resins are now sold in do-it-yourself packages. It is a simple matter to combine the two reactive liquids in front of the class. Do it inside a large Pyrex glass container, for the increase in volume is surprising.

11 Oxidizing Agents

After considerable discussion of flammable fuels of every description and origin, we are about to change the subject. From time to time, the subject of oxygen has come up: as part of the atmosphere or the fire rectangle; as a pressurized or a liquefied gas. In passing, we commented on the existence of a number of OXIDIZING AGENTS. As the name implies, these are substances containing oxygen atoms which can be released. If this loosening of insecure chemical bonds takes place during a fire, it can have some interesting consequences. Now is the time for a more thorough look at these oxidizers.

A problem arises immediately. The hazards of oxidizing agents vary. Which of them should we include here, which should we reserve for another, more appropriate, time and place? For example, although ammonium perchlorate and hydrogen peroxide are part of this chapter, they could just as logically have been included under "explosives" or "unstables." The choice depends on the particular properties we wish to consider. Often, there is a thin line between an oxidizing agent and an explosive. As we shall see, every one of these oxidizers is capable of crossing this line in one way or another, of exploding under the right (wrong) circumstances.

This chapter deals with oxidizing agents, but not those which are also flammable. We have met one troublemaker like this already, cellulose nitrate, and there are others, such as the organic peroxides. (These are known to the fire service as S.O.B.'s, which undoubtedly stands for self-oxidizing burnables.) When the temperature increase caused by decomposition is added to such substances, all sides of the fire rectangle are present in one unstable package.

None of the oxidizing agents in this chapter is considered combustible. Just as a fuel must be oxidized before combustion takes place, these oxidizing agents must be fueled. Rather arbitrarily, we have selected three loosely related groups of oxidizers for investigation. Their names sound like a zoo of strange chemical animals, awaiting an opportunity to bite.

1. The chlorites, chlorates, and perchlorates.
2. Some "pers": the permanganates, persulfates, and peroxides.
3. The nitrates and nitrites.

How to Build an Oxidizer

When one of two words in the name of a substance changes, we know that we are speaking of two completely different chemical compounds or groups of compounds. No one can ignore the difference between "celluose *nitrate*" and "cellulose *acetate*," or even between "ORGANIC peroxide" and "INORGANIC peroxide." But, we have to look closely when identification hinges upon the change of a single interior letter. Notice the close resemblance between these names of oxidizing agents: sodium nitr*at*e and sodium nitr*it*e; or even worse, sodium chlor*id*e, sodium chlor*it*e, and sodium chlor*at*e. Does it really matter? Very much so. For, in those sodium compounds, a change of one letter marks the difference between the salt you sprinkle on your scrambled eggs and an oxidizing agent that can scramble you. All this, however, leads us back into chemistry.

Groups of atoms inside a molecule often tend to operate as a unit. This section is all about these "units," for it is within these clusters of atoms that available oxygen is waiting to make an ignition easier, to make a fire more intense, to make an explosion possible.*

Cellulose nitrate introduced us to the nitrate group: NO_3. A nitrogen atom can also combine with two oxygen atoms to form a nitrite: NO_2. Similarly, chlorites (ClO_2), chlorates (ClO_3), and perchlorates (ClO_4) have available oxygen. In the case of oxidizing agents, these groups of atoms are generally, but not always, attached to some metal: barium, calcium, lead, magnesium, potassium, strontium, silver, sodium, etc. Putting them together with sodium, for example, gives the following results:

Sodium Chloride:	$NaCl$
Sodium Chlorite:	$NaClO_2$

*These groups have several names. Sometimes they are called radicals; sometimes, because of their net electrical charge, they are known as ions, cations, or anions.

Sodium Chlorate:	$NaClO_3$
Sodium Perchlorate:	$NaClO_4$
Sodium Nitrite:	$NaNO_2$
Sodium Nitrate:	$NaNO_3$*

Common sense tells us that the more oxygen a group contains, the more it can release to a fire, and the greater its hazard should be. This is true only some of the time. A nitrate is more hazardous than the equivalent nitrite. Sodium chlorite is certainly more dangerous than sodium chloride, which is not an oxidizer at all, but common table salt. Yet sodium perchlorate, with four oxygen atoms, is more stable than sodium chlorate, which only has three. And most authorities consider sodium chlorite every bit as dangerous as sodium chlorate.

Chlorites, Chlorates, and Perchlorates

As always, our first concern is with identification. Although there are some variations in uses and properties (refer to Table 11–1), most of the chlorites, chlorates, and perchlorates look about alike. It is difficult to tell one of these water-soluble white powders or crystals from another. They all bear the same yellow DOT label with black lettering and symbol: oxidizer; or organic peroxide, if the product is organic. Organic peroxides will be discussed in Chapter 3 of *Explosive and Toxic Hazardous Materials*.

To repeat and emphasize a point, these salts will not burn. They give off oxygen, especially when heated, and thereby increase the combustion rate of any flammable material nearby. This phrase, "increase the combustion rate," is too easily passed over unless you have witnessed a demonstration of what it means. On a small scale, a cloth soaked in a chlorate solution, allowed to dry, and set afire is here one second, gone the next, in a burst of flame. On a large scale, a fire involving these agents becomes extremely dangerous. The speed of burning, the amount of heat produced, and the rate of fire spread are above all normal expectations, and explosions are possible. Many commercial explosives, as we shall see in *Explosive and Toxic Hazardous Materials,* contain an oxidizer. Some explosives, with a few refinements, are nothing more than an intimate mixture of a finely-divided fuel and an oxidizing agent. Chlorates are used in priming caps. Large percentages of potassium chlorate and ammonium perchlorate are in a high explosive called Cheddite. All by itself, ammonium perchlorate has already gone a long way down the road toward being an explosive. It requires only a little encouragement to finish its journey.

*The chemical suffixes *-ite* and *-ate* refer to groups which contain oxygen. An *-ate* has one more oxygen than an *-ite*. Incidentally, the word *chloride* has nothing to do with oxygen; it merely informs you that chlorine is present.

Chlorites, chlorates, and perchlorates are potentially explosive in contact with many combustibles, including sulfur, organic materials, and metal powders. These mixtures not only ignite very easily and burn intensely, but can explode if subjected to shock, friction, or a heat increase. Moreover, these oxidizing agents decompose at different temperatures. The decomposition may occur with explosive rapidity, or containers may burst simply because of the vapor pressure buildup caused by decomposition gases. Explosions also happen because of chemical reactivity with nonflammable materials, such as certain acids and ammonium salts. Broadly speaking, chlorates are considered a bit more unstable than perchlorates.

One fire involving sodium chlorate occurred in a chemical laboratory of a university. Five employees were removing approximately 450 pounds (204 kg) of a chlorate from the basement of a building where it had been stored for years. Authorities had stressed the need for careful handling. The chlorate, in a number of metal and wooden boxes, was being loaded on a platform truck when there was a sizzling sound, then flame and smoke from one of the cartons. Two explosions and a flash fire followed. Four persons were killed; three more, injured. A number of students, teachers, and fire fighters suffered from smoke inhalation.

From our discussion, it is easy to see why flammable materials should *never* be stored in the same area with oxidizing agents. Storage buildings should be of fire-resistant or noncombustible construction. Often overlooked is the obvious fact that wooden floors will burn, so will wooden pallets, and so will surrounding refuse. If oxidizing agents are spilled, they must be cleaned up promptly, carefully avoiding the use of a combustible sweeping compound, and disposed of carefully. Containers, whether glass bottles or metal drums, should be protected in storage.

The amount of chlorite, chlorate, or perchlorate involved in a fire dictates fire strategy. Sizable quantities require the use of large streams from a distance with fire fighters in explosion-protected positions. Be careful not to knock containers about with hose streams. Smothering agents such as carbon dioxide and steam are useless against a fire that has a nonatmospheric source of oxygen. If ammonium perchlorate is being threatened by a large fire, it may be necessary to evacuate the surrounding area. Since the gases of decomposition are toxic, fire fighters should be protected with self-contained masks. Several of these salts are also poisonous. (See Table 11–1.)

What's a "Per"?

When the name of a chemical compound leads off with the prefix *per-*, it means that its molecule has a distinctive structure. The molecule can be so saturated with an element that there is no more room for additions. For instance, perchloroethylene, the common dry cleaning fluid, and perchloromethane (carbon tetrachloride) are loaded with chlorine (Figure 11–1, page 273).

TABLE 11–1. PROPERTIES AND USES OF CHLORITES, CHLORATES, AND PERCHLORATES

	PROPERTIES	USES
Ammonium perchlorate NH_4ClO_4	Powerful oxidizer. Very sensitive to explosion when shocked, exposed to heat, or contaminated. Decomposes upon heating.	Explosives, fireworks, etching and engraving.
Barium chlorate $Ba(ClO_3)_2$	*Poisonous.* Gives off oxygen at 482°F. (250°C.) Melts at 777°F. (414°C.).	Pyrotechnics (green fire), dye fixing, explosives, manufacture of other chlorates.
Barium perchlorate $Ba(ClO_4)_2 \cdot 3H_2O$	*Poisonous.* Melts at 941°F. (505°C.).	Drying agent for gases.
Calcium chlorate $Ca(ClO_3)_2 \cdot 2H_2O$	When rapidly heated, melts at 212°F. (100°C.). Hygroscopic.	Photography, pyrotechnics, dusting powder to kill poison ivy.
Magnesium perchlorate $Mg(ClO_4)_2$	Decomposes when heated above 482°F. (250°C.). Dissolves in water and creates considerable heat. Skin and eye irritant.	Drying agent for gases.
Potassium chlorate $KClO_3$	*Poisonous.* Decomposes at 752°F. (400°C.), giving off oxygen. Melts at 695°F. (368°C.). Explodes with sulfuric acid contact.	Oxidizing agent, explosives, matches, fireworks, percussion caps, medicines, dyes, bleach.
Potassium perchlorate $KClO_4$	Decomposes at 752°F. (400°C.). Irritating to eyes and skin	Explosives, medicine, oxidizer, photography, pyrotechnics, fuses.
Sodium chlorate $NaClO_3$	*Poisonous.* Liberates oxygen at 572°F. (300°C.). Melts at 491°F. (255°C.).	Oxidizer, matches, explosives, leather tanning, weed killer, bleach.
Sodium chlorite $NaClO_2$	*Poisonous.* Decomposes with heat at 347°F. (175°C.). Forms explosive and poisonous gas in contact with mineral acids.	Improving taste of water, bleach for textiles, paper, oils, waxes, shellac, varnishes, straws.
Sodium perchlorate $NaClO_4$	Decomposes at 266°F. (130°C.). Melts at 900°F. (482°C.). Irritating to eyes and skin.	Explosives, jet fuels.
Strontium chlorate $Sr(ClO_3)_2$	Decomposes at 248°F. (120°C.).	Pyrotechnics (red fire).

Perchloric acid, $HClO_4$, has a molecule completely filled with oxygen. Many other oxidizers can also be identified by this prefix: perchlorates, perborates, persulfates, permanganates, and others. Representatives of these groups can be found in Table 11–1 and Table 11–2 (page 274).

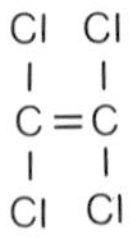

```
Cl  Cl              Cl
|   |               |
C = C          Cl - C - Cl
|   |               |
Cl  Cl              Cl
```

Perchloroethylene C_2Cl_4 | **Perchloromethane** CCl_4

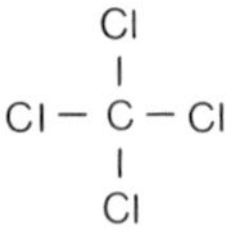

FIGURE 11–1. "Per" Compounds

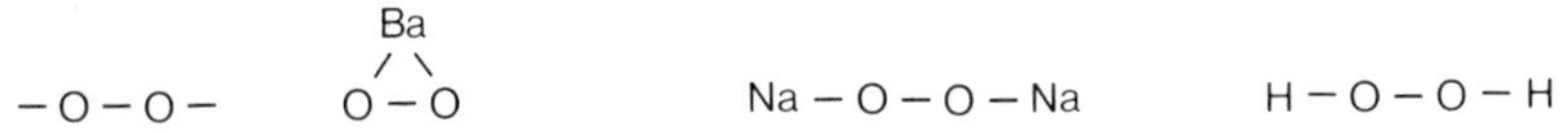

```
                Ba
                / \
- O - O -      O - O      Na - O - O - Na      H - O - O - H
```

Peroxy group | **Barium peroxide** BaO_2 | **Sodium peroxide** Na_2O_2 | **Hydrogen peroxide** H_2O_2

FIGURE 11–2. Inorganic Peroxides

Peroxides

Two oxygen atoms can hook together in the manner shown in Figure 11–2. Such a lineup is called a "peroxy group" and the compound which contains it, a "peroxide." When these oxygens are surrounded by groups of atoms containing carbon, the resulting compound is a highly flammable and unstable ORGANIC peroxide: benzoyl peroxide, for example, or methyl ethyl ketone peroxide. (See *Explosive and Toxic Hazardous Materials,* chapter on Unstable Materials.) If the peroxy group is joined to a metal such as sodium or barium, or if two hydrogen atoms are linked to the pair of oxygens, an INORGANIC peroxide is created.

Hydrogen Peroxide

The next time a lady you know decides to bleach her hair, don't mutter about the vanity of women. Take a good look instead at the bottle of peroxide she is using, for it is a remarkable liquid. Common household peroxide is a 3 percent solution. This means that 3 percent, by weight, of H_2O_2 has been dissolved in a large percentage of water. Hydrogen peroxide is completely soluble in water. At low concentrations, it forms a colorless, odorless liquid that is difficult to tell from pure water, except that it begins to show some of the characteristics which make it hazardous. As the lady is well aware, hydrogen peroxide can be used for a bleach. What she may not know is that the bleaching is caused by a controlled oxidation process that

TABLE 11–2. PERMANGANATES, PERSULFATES, PERBORATES (OXIDIZERS)

OXIDIZER	PROPERTIES	SHIPPING AND USES
Ammonium persulfate $NH_4S_2O_8$	White powder. Irritating to skin and mucous membranes. Yields toxic sulfur oxides when heated. Decomposes.	Shipped in containers from bottles to 300-pound (136-kg.) barrels. Uses: oxidizing agent, bleach, dyes. DOT yellow label.
Potassium permanganate $KMnO_4$ (Purple salt)	Dark purple crystals with a blue sheen. Decomposes at 464°F. (240°C.). Explodes in contact with organic materials, and with hydrogen peroxide either in solution or in dry state. Dangerous fire hazard. Ingestion of salt can cause death.	Shipped in containers from bottles to 500-pound (237-kg.) drums. Uses: wood preservative, deodorant, bleach, disinfectant (athlete's foot), dyes, absorbent for poison gas. DOT yellow label.
Potassium persulfate $K_2S_2O_8$ (Anthion)	White crystals. Decomposes and frees oxygen at 212°F. (100°C.) when dry. In solution, frees oxygen at room temperatures. Skin irritant. Yields toxic sulfur oxides when heated.	Shipped in containers from bottles to 300-pound (136-kg.) drums. Uses: bleach, oxidizing agent, photography, soap manufacture, antiseptic. DOT yellow label.
Sodium perborate $NaBO_2$	White crystals. Stable in cool, dry air, but will decompose and give off oxygen when heated or moist. Nonhazardous unless mixed with highly combustible or reactive compounds.	Shipped in cartons, kegs, barrels. Uses: dyes, bleaches, detergents, germicide, deodorant, neutralizing cold wave preparations. No shipping regulations.
Sodium permanganate $NaMnO_4$	Reddish-black crystals or powder. Decomposes. Dangerous fire hazard.	Shipped in containers from bottles to 300-pound (136-kg.) drums. Uses: oxidizing agent, disinfectant, antidote for certain poisons. DOT yellow label.

burns her hair to a lighter color. Hydrogen peroxide is a powerful oxidizer, but it is slow acting and therefore less damaging when it is used to bleach hair or such fabrics as wool, silk, linen, and fur.

How does it work? Hydrogen peroxide is nothing more than water with an extra oxygen atom jammed in.* If given the slightest chance, it will revert to water by decomposing.

*Technically, water can be called hydrogen oxide. The usefulness of this bit of information is limited, except to warn you that some pranksters have been known to include the location of large hydrogen oxide tanks on their industrial inspection reports, giggling as the fire inspectors dug into their chemical dictionaries looking for this dangerous-sounding compound.

$$2H_2O_2 \longrightarrow 2H_2O + O_2 + \text{HEAT}$$

This reaction has two interesting consequences. First, it is EXOTHERMIC (produces heat); second, it frees oxygen. But an overwhelming percentage of a 3 percent solution is water. The heat is easily absorbed and not too significant. Neither is the amount of oxygen set loose. Decomposition generally proceeds slowly, less than one percent a year under normal conditions, unless you leave the cap off the bottle. It would take a long time for household peroxide to turn back into pure water.

But if you hold a bottle of peroxide, the heat of your hand will soon release visible oxygen bubbles from the liquid. Your hand is about 20 degrees hotter than normal room temperatures. Hydrogen peroxide will decompose 1.5 times faster for each temperature rise of ten degrees Fahrenheit (5½ degrees Celsius). Also, there is a reason why peroxide bottles are invariably made of colored glass, dark blue or brown; decomposition of the liquid also increases when it is exposed to light.

Hydrogen peroxide is often used as a mild antiseptic, although it is relatively poor for the purpose. When put on a cut, it will foam. This foam is made of tiny oxygen bubbles. Foreign material in the blood causes an even more rapid breakdown of the peroxide than would be caused by exposure to body temperatures alone. As we shall discover, this "blood catalase," as it is called, is not unique in its ability to increase the decomposition rate of hydrogen peroxide.

What all of this means is that the rate of decomposition is variable and subject to change if H_2O_2 is contaminated, the temperature is increased, or even if light is allowed to strike the liquid. Despite this, hydrogen peroxide would still be considered a fairly innocuous liquid, an acceptable plaything, were it not for the fact that we have been describing its reactions in the lowest of concentrations.

We touched on something similar before when we discussed ammonia. At that time, we learned that industrial concentrations of ammonium hydroxide are far stronger and more dangerous than household ammonia. This is also true with hydrogen peroxide. To find out how a higher concentration of hydrogen peroxide accentuates its hazards, let us look at some of its common industrial strengths.

Toxicity

There is a warning inside most packages of hair lighteners: "This product contains ingredients which may cause skin irritation on certain individuals, and a preliminary test should first be made (on a patch of skin). This product must not be used on eyelashes or eyebrows. To do so may cause blindness."

When the concentration of H_2O_2 in water reaches 8 percent, a DOT yellow label: oxidizer, is required during shipment. In still higher concentrations, toxic hazards continue to increase. Vapors and mists become more and more irritating to the eyes, nose, and throat. The liquid can cause severe skin burns and blisters, especially in concentrations above 35 percent.

But you have an immediate antidote for this corrosiveness inside your fire hose. You only have to remember that hydrogen peroxide is *completely soluble* in water. Dangerous concentrations can be reduced to nontoxic levels when enough water is added. Fire fighters splashed by liquid peroxide should be thoroughly soaked by a hose stream. If your clothing absorbs peroxide, the corrosive action of the liquid will be accentuated by continual direct contact with your skin. Also, you are now wearing a highly oxygenated uniform. Therefore, you must undress and wash down immediately. Severe skin reactions are possible from exposures.

Taken internally, hydrogen peroxide is poisonous. It is extremely dangerous to the eyes. In the event of these types of contact, or if skin blistering occurs, or skin irritation is not corrected by flushing with water, get medical attention at once. Recommended first aid measures, after dilution with water, include the use of boric acid paste on flesh burns, a 3 percent solution of boric acid in the eyes.

A self-contained gas mask, of course, protects against some of these possibilities. Considering the nature of hydrogen peroxide, flame-proofed and nonabsorbent clothing is desirable. Workers around peroxide generally wear plastic aprons, gloves, and goggles. A safety shower is close at hand.

Peroxide Fire and Explosion Hazards

Hydrogen peroxide did not seem very dangerous when we talked about the reactions in a 3 percent solution, a lather of oxygen on a cut, little streams of oxygen bubbles in a bottle, and an increased heat of decomposition coming from contact with a hand—heat so easily absorbed by the large percentage of water. Industry puts peroxide to work as a bleach or a polymerization promoter, or in the manufacture of rayons, dyes, starches, glues, and gelatins. When we turn to industrial peroxide, we are dealing with storage concentrations of 35 and 50 percent; with shipments of 70 percent solutions that are diluted to usable strengths at the scene of action; or even, for special purposes, with 90 percent strengths or pure H_2O_2. Our quiet little reactions become as strong and dangerous as the peroxide itself. Several interwoven hazards must be countered by special storage and handling requirements.

Hydrogen peroxide can be a more powerful oxidizer than chlorine or pure oxygen. Like most oxidizing agents, it will increase the burning rate of an involved flammable. When in contact with a combustible material (and a liquid can get very close indeed), hydrogen peroxide can lower ignition temperature or even cause a flammable material to ignite spontaneously if such contact lasts any time. Explosive mixtures become possible as percentages become higher.

> Explosive mixtures may be formed with many organic materials, but with rare exceptions, an initial hydrogen peroxide concentration of more than 55 percent is needed. The shock sensitivity and energy is then of the same order as TNT. Explosions can be caused by mechanical shock, heat, or electrical discharge.*

Commerical hydrogen peroxide will decompose into oxygen and water vapor, producing heat. But we are not talking now of a small bottle. Industrial shipping containers and storage tanks may hold thousands of gallons. Since it is possible for one volume of decomposing liquid to yield a thousand volumes or more of gas, an explosion is inevitable from the force of gas pressure alone, unless the container has the proper emergency venting devices.

The heat production that was so trifling in a 3 percent solution is now something else. Up to a certain percentage, there is still enough water in a peroxide solution to absorb the heat of decomposition without boiling. But, at a concentration of 64.7 percent or above, this is no longer true. Although water itself is a decomposition product and more water is continually made by the decomposition process, the percentage of H_2O_2 begins to rise as the water boils away. (Hydrogen peroxide boils at a higher temperature than water: 286°F. [141°C.]). The process becomes cumulative. The very heat it is creating triggers the peroxide into an even faster reaction, almost a definition of an unstable material. Eventually, no liquid at all will be left. In a 90 percent concentration, the heat of the decomposition gases is over 1300°F. (704°C.); in a 100 percent concentration, their temperature is close to 1800°F. (982°C.), more than enough to ignite any combustible around, without considering the presence of oxygen.**

Heat is not the only cause of hydrogen peroxide's decomposition. Just as the blood catalase did, many contaminants will greatly increase the rate of decomposition—for instance, dust and dirt, or such metals as brass, bronze, copper, iron, lead, manganese, silver, and their salts. Notice that many common structural metals are included in this list. Permanganates and hypochlorites are often specifically mentioned as being very dangerous in this respect. There is still another possibility: although hydrogen peroxide is not flammable, a vapor concentration exceeding 26 percent in air is capable of violent decomposition, given a boost from a source of energy. This

*Factory Mutual, *Handbook of Industrial Loss Prevention*, 2nd Edition (New York: McGraw-Hill, 1967).

**These temperatures are calculated assuming no heat loss to the environment. Such a theoretical situation is called ADIABATIC.

vapor-air mixture exists naturally above an enclosed concentration of 74 percent hydrogen peroxide at normal temperatures and pressures. It can also exist above lower concentrations when temperatures and pressures are increased.

At concentrations of 90 percent and more, hydrogen peroxide becomes a heavy, syrupy liquid and a fire fighter's nightmare. The toxic hazard is extreme. Solutions, vapors, and mists are highly irritating. Explosive decomposition or explosive oxidation of combustibles at *ordinary temperatures* is possible. Above 200°F. (93°C.) an explosive decomposition of a 90 percent concentration is *to be expected*. Even if a mixture does not explode upon contact with its surroundings, the combination may subsequently detonate if it is shocked or catches fire. As we saw, the heat of decomposition is extremely high. Decomposition can be caused not only by contaminants, but also by light, temperature increase, agitation, or contact with rough surfaces. These liquids are workable at room temperatures only because they are handled gingerly and all contaminants and light are kept away. When used as rocket propellents, they are deliberately catalyzed into decomposition. The German V-1 used a hydrogen peroxide propellent during World War II, and our space program is still one of its major users. (See the chapter on Rocket Propellents in *Explosive and Toxic Hazardous Materials*.)

Shipping

What can be done to control such a demon? The answer starts with stringent shipping requirements. Special containers are always required for hydrogen peroxide solutions. Amber or dark blue glass bottles are only two of a number of containers that are especially designed for peroxide. DOT shipping regulations show careful consideration of the increasing hazards of higher concentrations and a full awareness that iron and steel are on the forbidden list of contaminants. The following summary is drawn from Section 173.266 of the regulations. (For a complele description, refer to the section itself.)

Hydrogen peroxide solutions in water must be packed in containers as follows:

Over 52%

1. Glass bottles, not over one-quart (one-liter) capacity, packed inside two metal containers, one vented at the top, one at the bottom. Cushioning material, noncombustible, at least 10 times the volume of the liquid, wet with at least 10 percent water and a stabilizing agent.

2. Aluminum drums with sealed and vented closures. Venting arrangements as approved by the Bureau of Explosives. There should be signs reading: KEEP THIS END UP or KEEP PLUG UP TO PREVENT SPILLAGE.

3. Tank cars must have venting arrangements as approved by the Bureau of Explosives.

52% or Less

1. Glass or earthenware containers of not more than one-gallon (four-liter) capacity, inside wooden boxes well cushioned with noncombustible material.

2. Carboys, single-trip drums, or 55-gallon drums, made of aluminum with vented closures.

8% to 37%

Vented polyethylene and glass carboys. Vented aluminum bottles not over 5 pounds or pints. Polyethylene or other plastic bottles not exceeding one pint capacity inside wooden boxes.

In addition to the above, concentrations between 8 and 10 percent may also be shipped in paraffin-lined wooden barrels.

Small amounts of hydrogen peroxide, 52 percent or less, are exempt from the above, but must be shipped in glass bottles inside metal cans. There are no exemptions when the concentration is over 52 percent.

Storage

Caution must be maintained after these containers reach their destination. Since drums are vented, do not stack them. (This also prevents a damaging fall.) Store and handle them in an upright position. To prevent contamination, store peroxides in the original containers, kept closed against contaminants or water escape that will increase the concentration of the remaining solution. Simply, store drums in a well-ventilated area where contaminants and sources of heat cannot disturb them. A detached, noncombustible building is ideal, although drums are often kept outside under a protective nonflammable canopy. Storage distances from occupied or important buildings depend again upon the strength of the solution. Below 64.7 percent, drums may be stored within 50 feet. At or above this percentage, 100 feet (30 meters) is the minimum. Limited storage of concentrations below 64.7 percent is allowed inside a main building if the storage room has walls of fire-resistive construction, nonflammable floors, drainage facilities, and adequate supplies of water for dilution of spills.

Special alloys of aluminum are used to build storage tanks. Piping is designed to prevent a backflow of contaminants or a trapping of peroxide. Even the gaskets are made of an inert material that will not catalyze H_2O_2. Dilution techniques and location of unloading areas are subject to strict regulations. In spite of all this, and dust filters too, decomposition can still start inside a large tank of highly concentrated peroxide. If it does, the importance of a separated location and adequate pressure relief devices on tanks becomes evident. Tanks are located outside in a position dictated by their size, strength of the solution they contain, and how the pressure inside them is to be relieved. Pressure relief devices are sized according to the tank they must vent, and they should be large enough to accommodate a severely contaminated solution. These devices include large manhole covers which vent at one psi, and rupture discs which let go at 15 psi (1 atmosphere). Equipped with manhole covers, tanks containing concentrations below 64.7 percent can be stored next to a noncombustible wall or 25 feet (7.6 meters) from a combustible wall. Approximate storage distances from an important building for a 2000-gallon (7.54-cubic-meter) peroxide tank, with and without pressure relief devices, are given in Table 11–3.

TABLE 11–3. STORAGE DISTANCE FOR A HYDROGEN PEROXIDE TANK OF 2000 GALLONS (7.54 M³)

PERCENTAGE OF SOLUTION	DISTANCE*		DISTANCE**	
	FEET	METERS	FEET	METERS
35	40	12	500	152
50	80	24	600	183
70	150	46	700	213
90	300	91	850	259
100	600	183	1200	366

*Distance for: 35% and 50% solutions equipped only with 15 psi (1 atmosphere) rupture discs; 70%, 90% and 100% solutions with manhole covers.

**Distance for: tanks not equipped with pressure relief. If tanks are barricaded, this distance can be cut in half. A tank farm should be located according to the distance required for largest tank.

Fire Fighting

When you are called into an emergency involving hydrogen peroxide, you need information—the correct answers to a number of questions. There may never be a time when you need technical advice more. The first question is, What has happened? Hydrogen peroxide may have spilled, formed an accidental mixture with a combustible, be exposed to fire, or the solution inside a tank or smaller container may be decomposing. Once you know what has happened, you can begin to think about a succession of further questions.

If there is a spill, how much water will be necessary to dilute it to a safe level? Where will all this liquid flow? Will people be exposed to a toxic hazard? What is the possibility of contamination? If the spill is continuing from piping, where are the remote shutoffs?

If a tank or a group of containers is threatened by, or involved in, a fire, what is its location in relation to important exposures? How are the tanks vented? Will the pressure relief devices be able to take care of the overpressure? Are the tanks aluminum rather than steel, the containers aluminum or plastic? Is it possible to keep them intact with a cooling spray stream?

What can be done about decomposition, particularly if a tank contains a concentration above 64.7 percent? Since emergency stabilization may be possible, who would know what stabilizers are needed and how to use them? Once again, are the containers properly vented, or is there a possibility of a pressure explosion? If the temperature inside a tank is 30 degrees above normal, or rising at the rate of one degree every 15 minutes, it probably means the situation within is deteriorating. Does the plant have a system that automatically monitors temperatures and sounds an alarm? Can this system also dump a tank into a safely diked location while simultaneously diluting the contents?

Armed with answers to these questions and to the others which will occur to you on the scene, some sort of workable plan can be put together.

To help, you must have a strong supply of water. Nothing can replace it to cool tanks and containers, to dilute and wash away spills, to protect exposures. With an oxidizer involved, it is the only efficient extinguishing agent at our command. However, if you still seem to be losing ground, evacuation may become necessary. The distance you withdraw depends upon your estimate of the situation, once again after consultation with plant personnel.

Inorganic Peroxides

When one of these double "O" peroxy groups joins with a metal, other inorganic peroxides are formed. (See Figure 11–2.) All are nonflammable oxidizing agents. We will focus our attention on the most common of them, SODIUM PEROXIDE, Na_2O_2, a yellowish-white powder which becomes yellower as its temperature rises. Hard lumps of this peroxide, called oxone, are sometimes encountered.

Sodium peroxide is quite versatile. It can be used to bleach paper, textiles, fats, oils, resins, animal products, bristles, straw, ivory, sponges, and feathers. It is also employed as a germicide, a deodorant, an oxidizing agent, in dyes, and for water purification. One of its more interesting functions is to regenerate the air in submarines, aircraft, and space vehicles.

Sodium peroxide will combine with carbon monoxide to form harmless soda ash, sodium carbonate.

$$Na_2O_2 + CO \longrightarrow Na_2CO_3$$

Sodium peroxide **Carbon monoxide** **Sodium carbonate**

Sodium peroxide will also react with the carbon dioxide in the air to form soda ash and fresh oxygen.

$$2Na_2O_2 + 2CO_2 \longrightarrow 2Na_2CO_3 + O_2$$

Sodium peroxide **Carbon dioxide** **Sodium carbonate** **Oxygen**

(A reaction like this, using potassium superoxide, KO_2, is put to work in Chemox and other oxygen-generating gas masks.)

Inorganic peroxides, built around one of the ALKALINE EARTH metals (magnesium, calcium, strontium, and barium) are considered to be less water-reactive than those built around the ALKALI metals (lithium, potassium, and sodium). Sodium peroxide reacts vigorously with water. Depending upon the ratio of water to peroxide, the reaction can become explosively rapid, releasing oxygen and a great deal of heat. The heat of reaction, especially in the presence of oxygen, is more than enough to ignite most nearby combustibles. A mixture of sodium peroxide and a flammable material can either be explosive or easily ignited by a relatively small source of heat, even by friction. The rate of combustion is very high. These peroxides must be kept away from flammable liquids; immediate ignition may take place, especially in the presence of water.

Sodium peroxide is shipped in a wide variety of containers: glass bottles, fiber cans or boxes, wooden kegs, boxes or barrels, ranging in weight up to 400 pounds (182 kg). Some of these containers are combustible; some of them can be broken. For obvious reasons, spilled peroxide must be cleaned up promptly. In storage, it should be separated from flammable materials, and protected from damage or moisture.

Fire-fighting procedures depend upon circumstances. A very small fire involving alkali metal peroxides can be isolated by dry chemical or even drowned if enough water is used. The heat of reaction is absorbed by the flooding. When larger amounts of peroxide are involved, water simply

TABLE 11–4. INORGANIC PEROXIDES

	PROPERTIES	SHIPPING AND USES
Barium peroxide BaO_2 (Barium dioxide; Barium superoxide)	Grayish-white powder. *Poisonous.* Avoid breathing dust and contact with skin. Decomposes slowly in cold water, more rapidly in hot water. Above 1100°F. (593°C.) it will decompose into oxygen and barium oxide, which is caustic and also poisonous. Melts at 842°F. (450°C.). Oxidizing agent.	Shipped in bottles, boxes, kegs, drums, barrels, multi-walled paper sacks. Uses: manufacture of oxygen and hydrogen peroxide, glass decolorizer, bleach, tracer bullets, to oxygenate water.
Calcium peroxide CaO_2	White or yellowish, odorless powder. Decomposes at 527°F. (275°C.). Practically insoluble in water. Oxidizing agent.	Shipped in drums up to 200 pounds (91 kg.). Uses: seed disinfectant, dentifrices, medicine.
Lithium peroxide Li_2O_2	White powder or yellow grains. Decomposes in water, giving off oxygen and heat. Water solutions at ordinary temperatures can be decomposed.	Uses: bleach, source of oxygen.
Magnesium peroxide MgO_2	White odorless powder. Gradually decomposed by water with release of oxygen. Oxidizing agent.	Shipped in drums up to 200 pounds (91 kg.). Uses: bleach, oxidizer, medicine.
Potassium peroxide K_2O_2	Yellow mass. Decomposes in water, giving off oxygen and quantities of heat. *Caustic.* Does not burn or explode alone, but when mixed with combustibles, these mixtures are explosive or ignite easily. See discussion on sodium peroxide in text.	Shipped in glass bottles, cans, boxes, kegs, barrels, metal drums. Uses: oxidizing agent, bleach, oxygen-generating masks such as the Chemox.
Strontium peroxide SrO_2 (Strontium dioxide)	White odorless powder. Decomposes in hot water or when heated. Mixtures with combustibles ignite easily or can explode. Oxidizing agent.	Shipped in drums up to 500 pounds (227 kg.). Uses: bleach, medicine, fireworks.

cannot be used on the material itself, but can wet down exposures and limit the spread of the fire.

Finally, sodium peroxide is caustic itself and will react with wet skin to form heat and a corrosive alkali; either can give you a serious burn. Avoid getting it on your skin or in your eyes, or breathing the dust. Wear full protective clothing. See Table 11–4 (page 283) for the toxic and reactive properties of other peroxides.

Nitrates

Sodium Nitrate

AIA Special Interest Bulletin 164 gives the following case history.

> The periodic occurrence of large loss fires involving buildings occupied for the handling and storage of sodium nitrate indicates a need for a review of information concerning the material. The fires have caused serious injuries and loss of life as well as heavy damage to property. They have been characterized by combustible building construction, large fire areas, combustible storage containers, rapid spread of flame, intense heat, and explosions. . . .
>
> In March, 1952, at Savannah, Georgia, a fire of unknown origin destroyed a combustible warehouse containing 6,000 tons of sodium nitrate. The fire spread so rapidly that sections of the building collapsed before the firemen arrived. The property damage amounted to approximately $560,000. In March, 1953, at Wilmington, North Carolina, a fire destroyed a large port terminal property containing 25,000 tons of sodium nitrate. The construction was of brick and wood and the area involved in the fire covered 160,000 square feet. Total property damage was estimated at $5,600,000.
>
> Several warehouses of combustible construction had been provided with automatic sprinkler protection which proved to be inadequate for sodium nitrate occupancy. In one instance, sections of the sprinkler piping were entirely consumed when they dropped into the molten sodium nitrate. In spite of this discouraging record, there is good reason to believe this material can be handled and stored safely and efficiently under proper conditions.

Fire fighters believe in shorter hours, more pay, and the efficiency of sprinkler systems. The thought of them melting in a fire is almost as upsetting as having the NFPA endorse the use of Molotov cocktails. Certainly, such a failure emphasizes the "need for a review of information" concerning sodium nitrate.

Continuing our practice of presenting chemistry in digestible amounts, let us comment quickly that an acid will react with a base to form a salt. When this acid is NITRIC, the salt formed is called NITRATE.* Economically, nitrates

---

*An acid ending in *-ic* always forms a salt ending in *-ate; -ous* acids form *-ite* salts: nitr*ic* to nitr*ate*, nitr*ous* to nitr*ite*, etc.

are the most important of the oxidizing agents. The most widely used of them is one variously called caliche, soda niter, Chilean saltpeter, or sodium nitrate. The names reflect the fact that it is not only produced synthetically, but mined in several areas of the world, including South America. It is easy to understand why tons of it can be piled up in a single warehouse when we look over a list of its uses: to manufacture other nitrates, sulfuric and nitric acids, glass, pyrotechnics, medicines, matches, some forms of dynamite and military explosives, dyes, and food preservatives; or as a fertilizer, an oxidizing material, a brine refrigerant, and, as we shall see shortly, one of the salts in molten baths.

With the exception of ammonium nitrate and nitrite, cases unto themselves, most of the nitrate and nitrite salts have similar properties, varying only in some details. We will discuss sodium nitrate, as typical, taking note of any exceptions. Table 11–5 lists many of the nitrate salts.

(Although sodium nitrite looks a little different, having slightly yellowish crystals instead of the pure white or colorless crystals of the nitrate, it is shipped under the same DOT yellow label and subject to the same storage regulations. In any event, it slowly oxidizes to sodium nitrate when exposed to air. After a time, its properties will be identical.)

What makes sodium nitrate troublesome is that it is an oxidizer, it is hygroscopic, it decomposes, and it melts. These four phrases summarize its properties. Let us see why each is important to us and what we can do about them.

Sodium nitrate is an oxidizing agent. When heated, it gives off oxygen. What will happen when it meets a flammable material depends on a large number of variables: How much of each is involved? What kind of flammable? How complete is the mixture? How hot is it? Violently rapid combustion is possible. So is an explosion. Certainly, the degree of flammability will be increased, and ignition will be made easier. However, nitrate-flammable mixtures generally require an ignition source before they burn or explode. This is not always the case when a chlorate or a high concentration of hydrogen peroxide is mixed with a combustible.

In any event, why risk finding out if an ignition source is needed? Far better to have all nitrates stored in a cool, dry place, entirely separated from combustibles. But, many buildings have wooden floors or favor storage on wood pallets. Sodium nitrate, being hygroscopic, will absorb moisture either from the air or from materials it contacts. This property has an interesting effect on wood. The nitrate dries the wood, making it crack and curl up into a series of tiny splinters. If these wood splinters ignite, fires of amazing intensity and rapid spread are possible because they are fanned by the presence of an oxidizing agent. Splintering is not confined to floors or pallets. Nitrate dust, settling on wooden ledges or beams, will convert all wooden exposures into several million wooden slivers, all awaiting

TABLE 11–5. PROPERTIES OF NITRATES

	PROPERTIES	USES
Barium nitrate (Nitrobarite) $Ba(NO_3)_2$	Lustrous white crystals. *Poisonous.* May be lethal if swallowed. Melts at 1097°F. (592°C.).	Pyrotechnics (gives a green flame), explosives.
Calcium nitrate (Norway Saltpeter) $Ca(NO_3)_2$	Colorless granules. Evolves heat when it dissolves in water. Decomposes at 270°F. (.132°C.). Melts between 44°F. (7°C.) and 1040°F. (560°C.), depending upon amount of water in the molecule.	Explosives, pyrotechnics, matches, radio tubes.
Cobaltous nitrate $Co(NO_3)_2$	Red crystals. Melts at 133°F. (56°C.). Decomposes at 165°F. (74°C.).	Inks, cobalt compounds and pigments, decorating stoneware.
Copper nitrate $Cu(NO_3)_2$	Blue crystals. *Poisonous.* Melts in range between 79°F. (26°C.) and 238°F. (114°C.). Decomposes at 338°F. (170°C.).	Medicine, preparation of light-sensitive papers, dyes.
Lead nitrate $Pb(NO_3)_2$	White crystals. *Poisonous.* Decomposes at 878°F. (470°C.).	Lead salts, medicines, paints, matches, special explosives.
Lithium nitrate $LiNO_3$	Colorless powder. Melts at 491°F. (255°C.).	Low temperature salt baths, heat exchange medium, ceramics.
Magnesium nitrate $Mg(NO_3)_2$	White powder. Melts at 203°F. (95°C.). Decomposes at 626°F. (136°C.).	Pyrotechnics.
Nickel nitrate $Ni(NO_3)_2$	Green crystals. Melts at 133°F. (56°C.). Boils at 277°F. (136°C.).	Nickel plating, brown ceramic colors, nickel catalysts.
Potassium nitrate (Saltpeter, Niter) KNO_3	White crystals or powder. Melts at 631°F. (333°C.). Decomposes at 752°F. (400°C.).	Pyrotechnics, salt baths, pickling meat, glass manufacture, treating tobacco to make it burn evenly.
Silver nitrate $AgNO_3$	Colorless to white, large or small crystals, which become grayish to black on exposure to light, or in the presence of organic materials. *Poisonous and corrosive.* May cause burns, dangerous to eyes. Wear protective equipment. Melting point at 414°F. (212°C.). Decomposes at 831°F. (444°C.).	Photography, hair dyeing, silver plating, silvering of mirrors, indelible inks, silver salts, medicine. Shipped in amber bottles up to 200 ounces (5.7 kg.). *MCA* warning label.

TABLE 11–5. PROPERTIES OF NITRATES (CONT.)

	PROPERTIES	USES
Strontium nitrate $Sr(NO_3)_2$	White crystals. Melts at 1058°F. (570°C.).	Marine signals, matches, railroad flares (red fire).
Thorium nitrate $Th(NO_3)_4$	White crystals or solid. Decomposes at 932°F. (500°C.).	Medicine, fluorine detection.
Uranyl nitrate (Uranium nitrate) $UO_2(NO_3)_2$	Yellow crystals. *Mildly radioactive,* protect personnel. *Dust can be toxic.* In ether solution, do not allow to stand in sunlight. Melts at 141°F. (61°C.). Boils at 244°F. (118°C.).	Photography, uranium glazes, medicine, uranium extraction.
Zinc nitrate $Zn(NO_3)_2$	Colorless lumps or crystals. Melts at 97°F. (36°C.). Boils at 268°F. (131°C.)	Medicine, dyeing.

ignition.* Even if the building is noncombustible, sodium nitrate is shipped in a wide variety of flammable containers, ranging from fiber drums to multi-walled paper bags which may or may not be moisture-proofed. If the containers become wet, the nitrate may partially melt and impregnate them. Upon drying, the containers become highly flammable. For all these reasons, noncombustible storage bins are recommended for bulk nitrate shipments. Shipping bags and barrels are supposed to be thoroughly washed or destroyed after use. Unfortunately, this safety precaution is usually not observed.

Like other nitrates, sodium nitrate will decompose. When decomposition starts, at 716°F. (380°C.), sodium nitrate begins to release the same insidious, delayed-action group of colorless to tan to orange to brownish gases as cellulose nitrate, the nitrogen oxides. These gases will poison you before you are aware of it. You must protect yourself in any nitrate fire with a self-contained gas mask. Table 11–5 gives the decomposition temperatures of other nitrates.

Sodium nitrate melts at 586°F. (308°C.), well within ordinary fire temperature range. A molten oxidizing agent, loose in a fire situation, is something to be feared. It can easily ignite combustibles as it runs around with its load of oxygen, looking for trouble. If we pour water on it, or if water and molten salt flow about until they meet, steam explosions will occur. The intensity of the fire, fed as it may be by finely divided splinters of wood and generous supplies of oxygen, is overwhelming. No wonder

*Other nitrates may be less hygroscopic. Potassium nitrate is one example. Plant protection hose lines can become defective when exposed to a dusty nitrate atmosphere for a time. Consider this during inspections.

sodium nitrate fires have been so destructive; no wonder a sprinkler system occasionally melts.

Quick detection and action are all-important in nitrate fires. Without a fast knockdown, extinguishment may be impossible. Watchmen or automatic alarms are essential, along with a sprinkler system and a fire department that can apply enough water to put out the fire in a hurry before molten salt forms. Once a nitrate melts in large quantities, using water will be dangerous at close quarters, and no other extinguisher is going to be much better.

As a wonderful example of a positive viewpoint, however, consider this from the AIA.

> In one fire, the wooden building was quickly destroyed because the water supply was exhausted in a short time. [But] no explosions were reported and good salvage was obtained on the sodium nitrate.

Salt Baths

Salt baths come in three temperature ranges with each range particularly suited for heat-treating a specific metal. Various chlorides are used in high temperature baths, between 1100°F. and 2450°F. (593°C. and 1343°C.); cyanide salts are found in medium temperature baths between 1000°F. and 1750°F. (538°C. and 954°C.). We are concerned here with the so-called LOW TEMPERATURE BATHS that operate in a range between 500°F. and 1100°F. (260°C. and 593°C.). These baths often contain molten nitrates, such as a combination of potassium nitrate and sodium nitrate.* Explosives are not uncommon in salt baths for several reasons: overheating, reaction with surroundings, reaction with metals, and trapped water or air. (The quotations below are from AIA Research Report 2.)

Overheating Whether because of failure of temperature controls or because of a mistaken setting, a salt bath can overheat. Nitrates start to decompose at temperatures around 750°F. (399°C.). For a while, the decomposition moves along slowly. At 1100°F. (593°C.) it speeds up, releasing quantities of the nitrogen oxide gases. Above 1300°F. (704°C.), violent decomposition is possible. Normally, operating temperatures of nitrate baths should be below 1000°F. to 1100°F. (538°C. to 593°C.). For example:

> The explosion took place in a building housing a molten salt bath for heat treating aluminum alloy. For several hours the bath was overheated, although the night watchman attempted to reduce the flow

*Although sodium nitrate melts at 586°F. (308°C.), and potassium nitrate at 631°F. (333°C.), a mixture of the two melts at 426°F. (219°C.).

> of gas to the burners. It was then noted that the molten nitrate was foaming and running over the floor. Before proper action could be taken, the bath exploded with such violence that one steel section, weighing approximately 650 pounds [295 kg] was hurled a distance of 1000 feet [305 meters].*

Another explosion and fire, caused by overheating, took the lives of two employees and injured sixteen others. The bath normally operated at a temperature of 960°F. to 1000°F. (516°C. to 538°C.). It was reported that the bath reached a temperature of 1300°F. to 1350°F. (704°C. to 732°C.) for a period of time directly preceding the explosion. Loss: $300,000.

Reaction with Surroundings These molten salts are oxidizing agents in a most dangerous condition. Careless storage of reserve stocks of nitrate, chemical reactions with a combustible within the bath, or even a reaction with the metal of the container itself are frequent causes of trouble. And, if the nitrate gets loose, the consequences can be disastrous.

One fire report speaks of the "molten nitrate scattered about after an explosion, setting fire to surrounding combustibles." Another report mentions how a welded seam was ruptured, "which allowed the molten nitrate to react violently with the accumulated carbon in the furnace." After another explosion, an analysis of the salt residue "indicated that one can of cyanide had been mistakenly added with the nitrates to the nitrate bath. The combination of the salts caused the explosion as the temperature was increased." Even this is possible: "sections of the steel plate of the salt bath container were reduced from ½″ to ⅛″ by the rapid oxidation of the salts."

Reaction with Metals We already know how greedy magnesium is for oxygen. Imagine what happens if magnesium is introduced into a nitrate salt bath that can supply all the oxygen it hungers for. "An explosion and fire occurred in a plant which was heat treating aluminum alloys. Magnesium alloy castings were accidentally dipped into the nitrate bath which was heated to a temperature of approximately 1000°F. [538°C.]. An explosive reaction followed and the plant and equipment were almost completely demolished."

Aluminum itself has caused explosions, either in overheated baths or by a thermite reaction, caused when particles of aluminum react with the iron oxide sludge on the bottom of the salt container.

Trapped Water or Air This final group of reactions interests us most. Salt baths have exploded because of trapped air or because water has been brought into the situation. Steam explosions have taken place when water comes in contact with the molten salt "either as a carry-over from a

*AIA Research Report 2. Quotations following are from the same source.

preliminary cleaning bath, [or from] overheated service piping, leaky roofs, operation of automatic sprinklers, and liquid foods placed on ledges near the baths for warming.''

As fire fighters, consider the following case:

> A fire occurred in the heat treating section of a large metal working plant which resulted in an explosion. The fire department was partly responsible for this explosion by carelessly deluging the heat treating section with streams of water. The water caused a tremendous steam explosion when it suddenly entered one of the molten salt baths. The steam pressure created was sufficient to destroy the structure and the loss was over $175,000.

Consequently, the use of sprinklers around nitrate salt baths has had considerable scrutiny. The results of recent tests were summarized this way:

> Sprinkler discharge at different pressures was directed onto typical molten salt baths without explosive results and without spattering of any significance from the standpoint of personal injury and property damage. In the absence of sprinklers, a serious fire must ultimately be fought with hose streams; there is record of personnel being driven from the building by spattering when a solid hose stream was accidentally directed into a salt bath. Hose streams cause more severe spattering than does the relatively gentle discharge from sprinklers. A non-combustible hood may be installed over the baths to prevent sprinkler water from contacting them directly.*

''Spattering'' in the preceding quotation equals ''steam explosion.''

Emergency Procedures When called in because of an overheated salt bath, make sure the heat supply to the furnace has been shut off and all work removed from the bath. If the temperature is still rising, get ready for an explosion.

> Fires usually result from spillage and leaks. They spread fast and are difficult to control. Dry sand, which should be stored in nearby buckets and bins, can confine and prevent the spread of the escaped melt. Carbon dioxide and approved dry powder type extinguishers may be used to extinguish burning carbonaceous material around the immediate vicinity of the salt bath. No vaporizing liquid (carbon tetrachloride), water, foam, or other aqueous extinguishing agents

*Factory Mutual, *Handbook of Industrial Loss Prevention,* 2nd Edition (New York: McGraw-Hill, 1967).

should be permitted in fighting molten salt fires. While it is obvious that the addition of water to a molten salt bath will in all probability cause formation of dangerous steam pockets and an explosion, precautionary warning notices (CHEMICAL BATH—USE NO WATER!—EXPLOSION DANGER!) should be placed in prominent locations to prevent mistakes. All persons in the neighborhood of baths should be instructed as to the dangerous practice of using water and the correct use of sand.

Cautioning placards to warn firemen outside the factory of the presence of nitrate baths should be displayed at the entrances of the plant. The management should invite the officers of the nearby fire companies to inspect and become thoroughly acquainted with the location of these baths.

One last word of caution: Besides the obvious fact that we will be dealing with salts hot enough to destroy flesh, do not forget the danger from breathing the nitrogen oxides. If a uniform somehow becomes impregnated with a nitrate salt, remember these are oxidizing agents. If your clothing catches fire, a severe, perhaps fatal, burn can occur. Change clothes at once.

QUESTIONS ON CHAPTER 11

1. Chlorites, chlorates, and perchlorates are capable of exploding during a fire. Name three ways this can happen.
2. What procedures may be necessary if ammonium perchlorate is threatened by a large fire?
3. How can the decomposition rate of hydrogen peroxide be increased?
4. At what percentage is a DOT yellow label, oxidizer, required on hydrogen peroxide solutions?
5. What percentage of hydrogen peroxide is generally required to form an explosive mixture with organic materials?
6. Why are concentrations of hydrogen peroxide of 64.7 percent and above so dangerous?
7. Name five materials which will greatly increase the decomposition rate of H_2O_2.
8. What reactions occur when sodium peroxide is contacted by water?
9. Sodium nitrate is hygroscopic. What does this mean, and why is it important to fire fighters?
10. What can happen when sodium nitrate melts in a fire?
11. Discuss the use of hose streams around molten salt baths.
12. Discuss the use of a sprinkler system around molten salt baths.

13. Why should empty bags and barrels which contained sodium nitrate be washed or destroyed after use?
14. Why is stacking hydrogen peroxide containers in storage a bad practice?
15. Discuss the first aid measures for a fireman splashed by hydrogen peroxide.

BIBLIOGRAPHY

American Insurance Association:

Research Report 2, ''Potential Hazards in Molten Salt Baths for Heat Treatment of Metals,'' 1954.

Special Interest Bulletin No. 164, ''Sodium Nitrate Storage.''

Factory Manual System:

Handbook of Industrial Loss Prevention, 2nd Edition, 1967.

VISUAL AIDS

A color film, ''Chemical Booby Traps,'' produced by General Electric, has a section on the effect of oxidizers on combustible materials. There are also two films for a class interested in the chemistry of the situation: ''Acids, Bases, and Salts'', 20 minutes, color, and ''Oxidation and Reduction'', 10 minutes, color. These films can be obtained from a film library that is oriented toward college-level scientific visual aids. Your local college can assist you.

DEMONSTRATIONS

Fire from Ice: Both the water reactivity of sodium peroxide and the accelerated burning rate given a combustible by an oxidizing agent can be demonstrated rather dramatically. You will need some excelsior or shredded paper, a small amount of sodium peroxide, an ice cube, and a noncombustible container. You can make quite a production of this demonstration. Work outdoors. Build a bed of excelsior on the bottom of your container. Sprinkle a couple of tablespoons of sodium peroxide on the excelsior. Put the ice cube on top of the peroxide. The cold water slows down the reaction enough to allow a getaway. Within a minute or less, enough water has melted to activate the peroxide which will then ignite the paper through the heat of reaction. Keep the class upwind. Sometimes the paper ignites with a pop.

The reactions in a bottle of household peroxide, described in the text, can easily be demonstrated.

Index

D

T

Z